ENVIRONMENTAL REMEDIATION TECHNOLOGIES, REGULATIONS AND SAFETY

WATER POLLUTION CONTROLS

EFFLUENT GUIDELINES, TOTAL MAXIMUM DAILY LOADS (TMDLS) AND STORMWATER PERMITS

Environmental Remediation Technologies, Regulations and Safety

Additional books in this series can be found on Nova's website under the Series tab.

Additional E-books in this series can be found on Nova's website under the E-book tab.

Water Resource Planning, Development and Management

Additional books in this series can be found on Nova's website under the Series tab.

Additional E-books in this series can be found on Nova's website under the E-book tab.

ENVIRONMENTAL REMEDIATION TECHNOLOGIES, REGULATIONS AND SAFETY

WATER POLLUTION CONTROLS

EFFLUENT GUIDELINES, TOTAL MAXIMUM DAILY LOADS (TMDLS) AND STORMWATER PERMITS

JULIA CRAWFORD
EDITOR

nova publishers
New York

For permission to use material from this book please contact us:
Telephone 631-231-7269; Fax 631-231-8175
Web Site: http://www.novapublishers.com

Library of Congress Cataloging-in-Publication Data

ISBN: 978-1-62417-441-4

Published by Nova Science Publishers, Inc. † New York

CONTENTS

PREFACE

Forty years after the Clean Water Act set a national goal of eliminating the discharge of pollutants into navigable U.S. waters, the Environmental Protection Agency (EPA) has made significant progress in reducing pollution from industrial facilities; nevertheless, pollution from these facilities continues to cause concern. EPA's actions to reduce this pollution have included establishing national technology-based regulations, or effluent guidelines, for separate industrial categories, such as petroleum refining, fertilizer manufacturing, coal mining, and metal finishing. Relatively few effluent guidelines have been revised or created in recent years and environmental advocacy groups continue to raise concerns because industrial facilities annually discharge hundreds of billions, and perhaps trillions of pounds of pollutants to U.S. waters. This book examines water pollution controls with a focus on effluent guidelines, total maximum daily loads and stormwater permits.

Chapter 1 – Under the Clean Water Act, EPA has made significant progress in reducing wastewater pollution from industrial facilities. EPA currently regulates 58 industrial categories, such as petroleum refining, fertilizer manufacturing, and coal mining, with technology-based regulations called effluent guidelines. Such guidelines are applied in permits to limit the pollutants that facilities may discharge. The Clean Water Act also calls for EPA to revise the guidelines when appropriate. EPA has done so, for example, to reflect advances in treatment technology or changes in industries.

Chapter 2 – Section 303(d) of the Clean Water Act (CWA) requires states to identify waters that are impaired by pollution, even after application of pollution controls. For those waters, states must establish a total maximum daily load (TMDL) of pollutants to ensure that water quality standards can be attained. A TMDL is both a quantitative assessment of pollution sources and pollutant reductions needed to restore and protect U.S. waters and a planning process for attaining water quality standards. Implementation of section 303(d) was dormant until states and the Environmental Protection Agency (EPA) were prodded by lawsuits. The program has been controversial, in part because of requirements and costs faced by states to implement this 40-year-old provision of the law, as well as industries, cities, farmers, and others who may be required to use new pollution controls to meet TMDL requirements.

Chapter 3 – The Environmental Protection Agency (EPA) and states are implementing a federally mandated program for controlling stormwater discharges from industrial facilities and municipalities. Large cities and most industry sources are subject to rules issued in 1990, and EPA issued permit rules to cover smaller cities and other industrial sources and construction sites in 1999. Because of the large number of affected sources and deadline changes that led to confusion, numerous questions have arisen about this program. Impacts and costs of the program's requirements, especially on cities, are a continuing concern.

In: Water Pollution Controls
Editor: Julia Crawford

ISBN: 978-1-62417-441-4

Chapter 1

WATER POLLUTION: EPA HAS IMPROVED ITS REVIEW OF EFFLUENT GUIDELINES BUT COULD BENEFIT FROM MORE INFORMATION ON TREATMENT TECHNOLOGIES*

United States Government Accountability Office

ABBREVIATIONS

EPA	Environmental Protection Agency
NPDES	National Pollutant Discharge Elimination System

WHY GAO DID THIS STUDY

Under the Clean Water Act, EPA has made significant progress in reducing wastewater pollution from industrial facilities. EPA currently regulates 58 industrial categories, such as petroleum refining, fertilizer

* This is an edited, reformatted and augmented version of The United States Government Accountability Office publication, Report to the Ranking Member, Subcommittee on Water Resources and Environment, Committee on Transportation and Infrastructure, House of Representatives, GAO-12-845, dated September 2012.

manufacturing, and coal mining, with technology-based regulations called effluent guidelines. Such guidelines are applied in permits to limit the pollutants that facilities may discharge. The Clean Water Act also calls for EPA to revise the guidelines when appropriate. EPA has done so, for example, to reflect advances in treatment technology or changes in industries.

GAO was asked to examine (1) the process EPA follows to screen and review industrial categories potentially needing new or revised guidelines and the results of that process from 2003 through 2010; (2) limitations to this process, if any, that could hinder EPA's effectiveness in advancing the goals of the Clean Water Act; and (3) EPA's actions to address any such limitations.

GAO analyzed the results of EPA's screening and review process from 2003 through 2010, surveyed state officials, and interviewed EPA officials and experts to obtain their views on EPA's process and its results.

What GAO Recommends

GAO is making recommendations to improve the effectiveness of EPA's effluent guidelines program by expanding its screening phase to better assess hazards and advances in treatment technology. EPA agreed with two recommendations in principle and said it is making progress on them, but said that one is not workable given current agency resources. GAO believes improvements can be made.

What GAO Found

The Environmental Protection Agency (EPA) uses a two-phase process to identify industrial categories potentially needing new or revised effluent guidelines to help reduce their pollutant discharges. EPA's 2002 draft *Strategy for National Clean Water Industrial Regulations* was the foundation for EPA's process. In the first, or "screening," phase, EPA uses data from two EPA databases to rank industrial categories according to the total toxicity of their wastewater. Using this ranking, public comments, and other considerations, EPA has identified relatively few industrial categories posing the highest hazard for the next, or "further review," phase. In this further review phase, EPA evaluates the categories to identify those that are appropriate for new or revised guidelines because treatment technologies are available to reduce pollutant discharges. Since 2003, EPA has regularly screened the 58 categories for which it has issued effluent guidelines, as well as some potential new industrial categories, and it has identified 12 categories for its further review phase. Of these 12 categories, EPA selected 3 for updated or new effluent guidelines. EPA chose not to set new guidelines for the others.

Limitations in EPA's screening phase may have led it to overlook some industrial categories that warrant further review for new or revised effluent guidelines.

Specifically, EPA has relied on limited hazard data that may have affected its ranking of industrial categories. Further, during its screening phase, EPA has not considered the availability of advanced treatment technologies for most industrial categories. Although its 2002 draft strategy recognized the importance of technology data, EPA has stated that such data were too difficult to obtain during the screening phase and, instead, considers them for the few categories that reach further review. Officials responsible for state water quality programs and experts on industrial discharges, however, identified categories they believe EPA should examine for new or updated guidelines to reflect changes in their industrial processes and treatment technology capabilities. According to some experts, consideration of treatment technologies is especially important for older effluent guidelines because changes are more likely to have occurred in either the industrial categories or the treatment technologies, making it possible that new, more advanced treatment technologies are available.

Recognizing the limitations of its hazard data and overall screening approach, EPA has begun revising its process but has not assessed other possible sources of information it could use to improve the screening phase. In 2012, EPA supplemented the hazard data used in screening with four new data sources.

EPA is also developing a regulation that, through electronic reporting, will increase the completeness and accuracy of its hazard data. In 2011, EPA also began to obtain recent treatment technology literature. According to EPA, the agency will expand on this work in 2013. Nonetheless, EPA has not thoroughly examined other usable sources of information on treatment technology, nor has it reassessed the role such information should take in its screening process. Without a more thorough and integrated screening approach that both uses improved hazard data and considers information on treatment technology, EPA cannot be certain that the effluent guidelines program reflects advances in the treatment technologies used to reduce pollutants in wastewater.

September 10, 2012

The Honorable Timothy H. Bishop
Ranking Member
Subcommittee on Water Resources and Environment
Committee on Transportation and Infrastructure
House of Representatives

Dear Mr. Bishop:

Forty years after the Clean Water Act set a national goal of eliminating the discharge of pollutants into navigable U.S. waters, the Environmental Protection Agency (EPA) has made significant progress in reducing pollution from industrial facilities; nevertheless, pollution from these facilities continues to cause concern.[1] EPA's actions to reduce this pollution have included establishing national technology-based regulations—or effluent guidelines—for separate industrial categories, such as petroleum refining, fertilizer manufacturing, coal mining, and metal finishing. EPA issued the vast majority of these regulations in the 1970s and 1980s and has revised most of them; revisions may range from changes in testing methods to establishment of more stringent standards. Relatively few effluent guidelines have been revised or created in recent years, however, and environmental advocacy groups continue to raise concerns because industrial facilities annually discharge hundreds of billions, and perhaps trillions of pounds of pollutants to U.S. waters. According to EPA, industrial pollutants may contribute, in part, to impaired water quality; harm aquatic life; and limit the ways in which people can safely use the nation's waters.

One of EPA's main responsibilities under the act is to regulate "point source" pollution—that is, pollution such as effluent or wastewater coming from a discrete point, such as a pipe from an industrial facility. The Clean Water Act directed EPA to establish effluent guidelines to achieve pollutant reductions using specific treatment technologies or changes in a facility's production processes. In establishing and revising effluent guidelines, EPA is to assess (1) the performance and availability of the best pollution control technologies or pollution prevention practices for an industrial category; (2) the economic achievability of those technologies; (3) non-water-quality environmental impacts, such as the energy required to reduce pollutants; and (4) other factors that the EPA Administrator deems appropriate, such as the risk posed by discharges. The legislative history of relevant provisions in the

Clean Water Act suggests that effluent guidelines were expected to be revised and made more stringent over time to reflect technological advances.

To carry out its effluent guidelines program, EPA develops regulations setting national effluent guidelines, and states generally implement the program by applying limits in permits that they issue to specific facilities. Under the National Pollutant Discharge Elimination System (NPDES) program, all facilities that discharge pollutants from any point source into U.S. waters are required to obtain a permit, typically from their state or EPA region. Under the Clean Water Act, EPA has authorized 46 states to issue NPDES permits and retains the authority to issue permits for the remaining 4 states: Idaho, Massachusetts, New Hampshire, and New Mexico.[2]

The Clean Water Act requires that after setting effluent guidelines, EPA is to annually review each existing effluent guideline—that is, guidelines for regulated industrial categories—to determine whether revising these guidelines would be appropriate. In addition, at least every 2 years, EPA is to identify industrial categories that do not have effluent guidelines but that discharge nontrivial amounts of toxic or certain other pollutants.[3] At least every 2 years, EPA is required to publish schedules for its annual review and revision of existing effluent guidelines and for promulgating effluent guidelines for any newly identified categories. The agency's intent is to issue a plan every year, with preliminary plans to be issued in odd years and final plans for effluent guidelines in even years. If EPA decides that an industrial category requires new or revised effluent guidelines, it generally establishes them through a regulatory process that involves proposing new effluent guidelines, obtaining public comment, making revisions, and publishing a final regulation.

Throughout much of the effluent guidelines program's history, EPA's schedule for issuing effluent guidelines has been driven by litigation and resulting consent decrees.[4] In 2002, following extensive consultation with an advisory task force formed in response to a 1992 consent decree, EPA issued a draft *Strategy for National Clean Water Industrial Regulations*, outlining a new process by which it planned to meet the requirement to review industries in the future to determine whether new or revised effluent guidelines were appropriate. The draft strategy calls for EPA to conduct an annual screening of industrial categories to consider (1) the risks the industrial categories pose to human health or the environment; (2) the availability of treatment technology or other approaches to reduce the risk; (3) the cost, performance, and affordability of the technology; and (4) implementation and efficiency considerations. EPA derived these screening factors in part from the statutory requirements for developing or revising effluent guidelines. Following screening with available information, the draft strategy calls for EPA to conduct a further review of selected categories. The further reviews may take 1 or more years to complete. EPA has not finalized or formally updated its draft strategy, although according to EPA officials, the draft has served in part as the basis for the agency's annual reviews of industrial categories after 2002.

As EPA's regulatory efforts have reduced pollutants from industrial point sources over the past several decades, the agency has placed greater emphasis on what is now the primary reason for impairment of the nation's waters, namely diffuse or nonpoint pollution, such as some agricultural runoff. In light of that change in emphasis and soon after issuing the draft strategy, EPA reduced staffing levels for the effluent guidelines program by about 40 percent, according to program officials. EPA issued its most recent effluent guidelines—for airport deicing, a previously unregulated industry—in May 2012. Before that, EPA's most recent revisions of existing effluent guidelines were for concentrated animal feeding operations in 2008 and construction and development in 2009.[5] Most effluent guidelines have not been revised since the 1980s or 1990s.

In this context, you asked us to review EPA's effluent guidelines program. This report examines (1) the process EPA follows to screen and review industrial categories potentially in need of new or revised effluent limitation guidelines and the results of that process from 2003 through 2010; (2) limitations to this screening and review process, if any, that could hinder the effectiveness of the effluent guidelines program in advancing the goals of the Clean Water Act; and (3) what actions EPA has taken or could take to address limitations, if any, that exist.

To address our objectives, we reviewed the Clean Water Act and relevant regulations, EPA's 2002 draft strategy, effluent guidelines program plans, and associated supporting documents. We also reviewed EPA's screening decisions for all industrial categories and its further reviews for the 12 industrial categories selected through screening from 2003 through 2010.[6] Our purpose was to identify those industries that EPA had only initially screened and those that received a further review, including an examination of available treatment technologies. We also documented the status of regulatory actions and other steps that EPA took for industries that it reviewed further. In addition, we interviewed officials in EPA's Engineering and Analysis Division to learn about the process the agency follows to screen and review industries potentially in need of new or revised effluent limitation guidelines. We then compared the steps specified in the draft strategy with the agency's current process for screening and reviewing industries for possible revised guidelines.

To better understand the steps in the current process as they apply to specific industrial categories, we conducted detailed interviews with EPA staff regarding 7 of the 12 industrial categories that EPA selected from 2003 onward for possible new or revised effluent guidelines. We chose 2003 because it was the year when EPA issued its first preliminary effluent guidelines plan after developing its 2002 draft strategy for screening and reviewing industries. We also conducted 17 interviews with 22 experts from academia, industry, nonprofit organizations, and state and local water quality agencies for their perspectives on EPA's effluent guidelines program. We selected these experts from a list of approximately 50 individuals identified from a variety of sources, including referrals from EPA, the Association of Clean Water Agencies, the National Association of Clean Water Agencies, and other experts; relevant academic literature; and litigation documents. Because we used a nonprobability sampling method to select experts, the results of our interviews with them cannot be generalized to all experts on the program, but the information derived from interviewing these experts provided illustrative observations and examples. We also surveyed the directors for water quality permits in the 46 states authorized to issue NPDES permits about the adequacy of current effluent guidelines; the results of our analysis are not generalizable to all industrial categories in all states. Using the results of the survey, we selected an industrial category that state officials said warranted revised effluent guidelines and interviewed state officials to learn more about the reasons for their views. We also interviewed EPA officials about their plans, if any, related to those industries. Appendixes I and II present a more detailed description of our scope and methodology. We conducted this performance audit from September 2011 to September 2012, in accordance with generally accepted government auditing standards. Those standards require that we plan and perform the audit to obtain sufficient, appropriate evidence to provide a reasonable basis for our findings and conclusions based on our audit objectives. We believe that the evidence obtained provides a reasonable basis for our findings and conclusions based on our audit objectives.

BACKGROUND

Tens of thousands of industrial facilities directly discharge wastewater into the waters of the United States and are subject to permit limits on their discharges, which for certain industries are determined by effluent guidelines set by EPA under the Clean Water Act. For certain industries, EPA issues a similar type of regulation—pretreatment standards— applicable to facilities that are indirect dischargers; that is, their effluent goes to wastewater treatment plants, which then discharge the collected and treated wastewater into a water body. To establish pollutant control limits for different pollutants in these guidelines or standards, EPA groups industrial facilities into categories that have similar products or services. To date, EPA has issued effluent guidelines or pretreatment standards for 58 industrial categories. EPA has issued effluent guidelines for 57 of the 58 categories and pretreatment standards for 35 of the 58 categories.[7] Table 1 lists industrial categories that are regulated by effluent guidelines and pretreatment standards. According to EPA, there are approximately 35,000 to 45,000 direct dischargers covered by effluent guidelines and about 10,000 facilities that discharge indirectly to wastewater treatment plants.

Table 1. Industrial Categories with Effluent Guidelines and Pretreatment Standards as of August 2012

Industrial category	Effluent guideline	Pretreatment standard
Airport deicing	X	
Aluminum forming	X	X
Asbestos manufacturing	X	
Battery manufacturing	X	X
Canned and preserved fruits and vegetables processing	X	X
Canned and preserved seafood processing	X	
Carbon black manufacturing	X	X
Cement manufacturing	X	
Centralized waste treatment	X	X
Coal mining	X	
Coil coating	X	X
Concentrated animal feeding operations	X	X
Concentrated aquatic animal production	X	
Construction and development	X	
Copper forming	X	X
Dairy products processing	X	
Electrical and electronic components	X	X
Electroplating	X	X
Explosives manufacturing	X	
Ferroalloy manufacturing	X	
Fertilizer manufacturing	X	X

Industrial category	Effluent guideline	Pretreatment standard
Glass manufacturing	X	X
Grain mills	X	X
Gum and wood chemicals manufacturing	X	
Hospital	X	
Ink formulating	X	X
Inorganic chemicals manufacturing	X	X
Iron and steel manufacturing	X	X
Landfills	X	
Leather tanning and finishing	X	X
Meat and poultry products	X	
Metal finishing	X	X
Metal molding and casting	X	X
Metal products and machinery	X	
Mineral mining and processing	X	
Nonferrous metals forming and metal powders	X	X
Nonferrous metals manufacturing	X	X
Oil and gas extraction	X	X
Ore mining and dressing	X	
Organic chemicals, plastics, and synthetic fibers	X	X
Paint formulating	X	X
Paving and roofing materials (tars and asphalt)	X	X
Pesticide chemicals	X	X
Petroleum refining	X	X
Pharmaceutical manufacturing	X	X
Phosphate manufacturing	X	
Photographic	X	
Plastics molding and forming	X	
Porcelain enameling	X	X
Pulp, paper, and paperboard	X	X
Rubber manufacturing	X	X
Soap and detergent manufacturing	X	X
Steam electric power generating	X	X
Sugar processing	X	
Textile mills	X	
Timber products processing	X	
Transportation equipment cleaning	X	X
Waste combustors	X	X
Total	57	35

Source: GAO analysis of EPA data.

National Pollutant Discharge Elimination System Permits

Before an industrial facility discharges pollutants, it must receive a permit that is to, at a minimum, incorporate any relevant pollutant limits from EPA's effluent guidelines. Where needed to protect water quality as determined by standards set by individual states, NPDES permits may include limits more stringent than the limits in the guidelines. NPDES permits for direct dischargers are issued by 1 of the 46 states authorized by EPA to issue them and by EPA elsewhere. Unlike direct dischargers, indirect dischargers, which do not discharge to surface waters, do not require an NPDES permit. Instead, an indirect discharger must meet EPA's national pretreatment standards and may have to meet additional pretreatment conditions imposed by its local wastewater treatment plant.[8] Under the national pretreatment standards and conditions, an indirect discharger is required to remove pollutants that may harm wastewater treatment plant operations or workers or, after treatment and discharge, cause violations of the wastewater treatment plant's permit. Figure 1 illustrates both types of facilities subject to regulation.

To get an NPDES permit, industrial facilities' owners—like any source discharging pollutants as a point source—must first submit an application that, among other things, provides information on their proposed discharges.

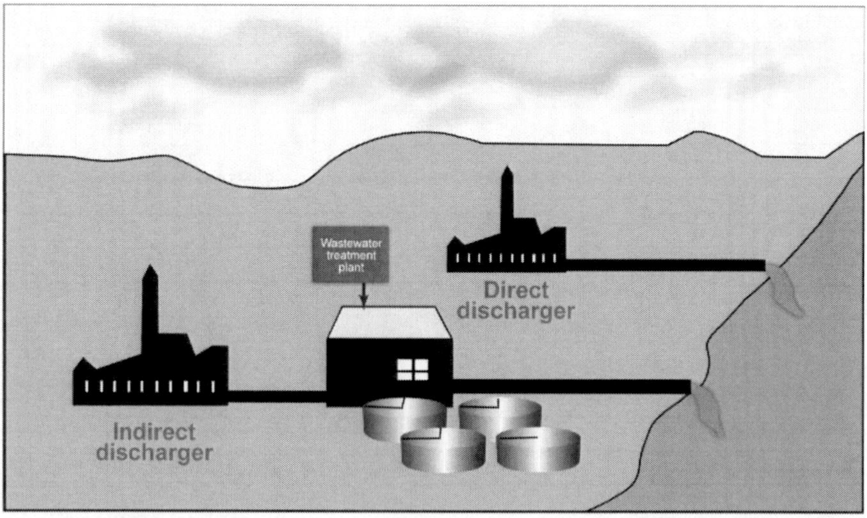

Sourc: EPA Documentation.

Figure 1. Industrial Facilities Subject to Regulation of Discharges.

Water quality officials in authorized states and EPA regional offices responsible for the NPDES program in the four nonauthorized states review these applications and determine the appropriate limits for the permits. Those limits may be technology-based effluent limits, water quality-based effluent limits, or a combination of both. Technology-based limits must stem from either effluent limitation guidelines, when applicable, or from the permit writer's best professional judgment when no applicable effluent limitation guidelines are available. Using best professional judgment, permit writers are to develop technology-based permit conditions on a case-by-case basis, considering all reasonably available and relevant information, as well as factors similar to those EPA uses in developing guidelines for national effluent limitations. A permit writer should also set water quality-based limits more stringent than technology-based limits if necessary to control pollutants that could cause or contribute to violation of a state's water quality standards. To support each permit, permit writers are supposed to develop a fact sheet, or similar documentation, briefly summarizing the key facts and significant factual, legal, methodological, and policy questions considered.[9] The fact sheet and supporting documentation also serve to explain to the facility, the public, and other interested parties the rationale and assumptions used in deriving the limitations in the permit.

Facilities with NPDES permits are required to monitor their discharges for the pollutants listed in their permits and to provide monitoring reports with their results to their permitting authority (the relevant state, tribal, or territorial agency authorized to issue NPDES permits or, in nonauthorized locations, EPA). For facilities designated by EPA regional administrators and the permitting authorities as major facilities, the permitting authorities are in turn required to transfer the monitoring report data to EPA headquarters. These reports, known as discharge monitoring reports, are transmitted electronically and stored in an electronic database or reported in documents and manually entered into the electronic database for use by EPA in reviewing permit compliance. [10] Permitting authorities are not required to report the discharge monitoring results from all remaining facilities, known as minor facilities, to EPA but may do so. According to EPA, there are about 6,700 major and 40,500 minor facilities covered by NPDES permits.

Facilities may also be required to report data to EPA's Toxics Release Inventory on their estimated wastewater discharges.[11] This inventory contains annual estimates of facilities' discharges of more than 650 toxic chemicals to the environment. One of the inventory's primary purposes is to inform communities about toxic chemical releases to the environment, showing data from a wide range of mining, utility, manufacturing, and other industries subject to the reporting requirements. As such, although the inventory is unrelated to the NPDES program, the Toxics Release Inventory contains estimated discharges of toxic pollutants for many NPDES-permitted facilities. Not all industrial categories covered by effluent guidelines—the oil and gas industrial category, for example—are necessarily required to report to the inventory.

Effluent Guidelines Program

Under the Clean Water Act, EPA must establish effluent guidelines for three categories of pollutants—conventional, toxic, and nonconventional pollutants—and several levels of treatment technology. As defined in EPA's regulations, conventional pollutants include biological oxygen demand,[12] total suspended solids,[13] fecal coliform bacteria,[14] oil and grease, and pH.[15] The Clean Water Act designates toxic pollutants as those chemicals listed in a key congressional committee report,[16] which contains 65 entries, including, arsenic, carbon tetrachloride, and mercury, as well as groups of pollutants, such as halomethanes.[17] Nonconventional pollutants are any pollutants not designated as a conventional or toxic pollutant; for example, EPA has developed limitations for such nonconventional pollutants as chemical oxygen demand,[18] total organic carbon, and the nutrients nitrogen and phosphorus.

The act authorizes EPA to establish effluent limits for these three pollutant categories according to several standards; the standards generally reflect increasing levels of treatment technologies. A treatment technology is any process or mechanism that helps remove pollutants from wastewater and can include filters or other separators, biological or bacteria-based removal, and chemical neutralization. Legislative history of the Clean Water Act describes the expectation of attaining higher levels of treatment through research and development of new production processes, modifications, replacement of obsolete plans and processes, and other improvements in technology, taking into account the cost of treatment. [19] Under the act, the effluent limits do not specify a particular technology to be used but instead set a performance level based on one or more particular existing treatment technologies. Individual facilities then have to meet the performance level set but can choose which technology they use to meet it.

Under the act, EPA was to issue initial guidelines for existing facilities on the basis of the "best practicable control technology currently available" for conventional, toxic, and nonconventional pollutants—guidelines to be achieved by 1977—followed by guidelines set on the basis of "best available technology economically achievable" for toxic and nonconventional pollutants and "best conventional pollutant control technology" for conventional pollutants.

Table 2. Standards for Effluent Guidelines for Direct Dischargers

Standard	Pollutants	Basis for treatment level	Entities subject to regulation
Best practicable technology currently available	Toxics, nonconventional, and conventional	The average of the best performances of facilities within the industry	Existing industrial facilities during the Clean Water Act's initial implementation phase (1977-89)
Best conventional pollutant control technology	Conventional	The most stringent technology option that passes tests as feasible and economically achievable	Existing industrial facilities, after 1989[a]
Best available technology economically achievable	Toxics and nonconventional	Level to be set with reference to the best performer in any industrial category and determined to be economically achievable for the category or subcategory	Existing industrial facilities, after 1989[b]
New source performance standards	Toxics, nonconventional, and conventional	The most stringent controls attainable through the application of the best demonstrated control technology that does not pose a barrier to entry	New industrial facilities

Source: GAO analysis.

[a] If EPA has not established an applicable best conventional technology effluent guideline, then the best practicable treatment effluent guideline still applies.

[b] If EPA has not established an applicable best available technology effluent guideline, then the best practicable treatment effluent guideline still applies.

The act also called for guidelines known as "new source performance standards," which would apply to new facilities starting operations after such standards were proposed. When permitting authorities develop a permit, they apply standards most appropriate to a given facility: For example, a new facility would receive a permit with limits reflecting the new source performance standards. Existing facilities would generally receive permits with limits reflecting the best conventional technology and best available technology, but where those standards have not been issued, permit limits would reflect best practical treatment. Table 2 shows the different levels of treatment established in the act and the category of pollutant to which they apply.

The Clean Water Act requires EPA to annually review all existing effluent guidelines and revise them if appropriate, and also to review existing effluent limitations at least every 5 years and revise them if appropriate.[20] The Water Quality Act of 1987 added two related requirements to EPA's reviews. First, EPA is to identify, every 2 years, potential candidates for new effluent guidelines, namely, industries that are discharging significant, or nontrivial, amounts of toxic or nonconventional pollutants that are not currently subject to effluent guidelines. Second, every 2 years beginning in 1988, EPA is required to publish a plan establishing a schedule for the annual review and revision of the effluent guidelines it has previously promulgated. In response to these two requirements, EPA published its first effluent guidelines program plan in 1990, which contained schedules for developing new and revised effluent guidelines for several industrial categories.

From the start of the effluent guidelines program in the early 1970s, EPA has faced considerable litigation, with industry challenging most of the industry-specific effluent guidelines. As the agency implemented the program, EPA also faced challenges from environmental groups over its failure to issue guidelines and the process EPA used to screen and review industrial categories. For example, the Natural Resources Defense Council, an environmental organization, brought two suits, each seeking to compel EPA to meet its duties to promulgate effluent limitations for listed toxic pollutants, among other actions. As a result, EPA operated under two key consent decrees establishing court-approved schedules for it to develop and issue effluent guidelines regulations. In addition, under one of the consent decrees, EPA established a task force that operated from 1992 through 2000 and advised the agency on various aspects of the effluent guidelines program. In particular, the task force issued several reports advising EPA on changes to its screening and review process for the effluent guidelines program and recommended that EPA hold a workshop to discuss improvements to the process.

In 2002, after considering the recommendations made by both the task force and the workshop, EPA developed an approach to guide its post-consent decree screening and review, issued in a document called *A Strategy for National Clean Water Industrial Regulations*. Under this draft strategy, EPA was to evaluate readily available data and stakeholder input to create an initial list of categories warranting further examination for potential effluent guidelines. The strategy identified the following four key factors for EPA to consider in deciding whether to revise existing effluent guidelines or to develop new ones:

- the extent to which pollutants remaining in an industrial category's discharge pose a substantial risk to human health or the environment;
- the availability of a treatment technology, process change, or pollution prevention alternative that can effectively reduce the pollutants and risk;
- the cost, performance, and affordability of the technology, process change, or pollution prevention measures relative to their benefits; and
- the extent to which existing effluent guidelines could be revised, for example, to eliminate inefficiencies or impediments to technological innovation or to promote innovative approaches.

The draft strategy also indicated that EPA would apply nearly identical factors to help determine whether it should issue effluent guidelines for industrial categories for which it had not yet done so. The document noted that EPA intended to revise and issue the strategy in early 2003, but EPA has chosen not to finalize it.[21] EPA officials stated that the agency made this choice because its implementation of the process was likely to evolve over time.

Since EPA issued its draft strategy, the agency has faced litigation challenging the use of technology in its screening process. In 2004, EPA was sued by Our Children's Earth, a nonprofit environmental organization, which alleged that EPA failed to consider technology-based factors during its annual review of industrial categories. On appeal, the Ninth Circuit Court decided in 2008 that the statute did not establish a mandatory duty for EPA to consider such factors. The court found that the statute's use of the phrase "if appropriate" indicated that decisions on whether to revise guidelines are discretionary but are also constrained by the statute's mandate as to what effluent guidelines regulations are to accomplish.22 Further, the court stated that the overall structure of the Clean Water Act strongly suggests that any review to determine whether revision of effluent guidelines is appropriate should contemplate technology-based factors.

EPA'S TWO-PHASE SCREENING AND REVIEW PROCESS HAS IDENTIFIED FEW INDUSTRIAL CATEGORIES FOR NEW OR REVISED EFFLUENT GUIDELINES

EPA uses a two-phase process to review industrial categories potentially in need of new or revised effluent guidelines; from 2003 through 2010, the agency identified few such categories. Since 2003, EPA has annually screened all industrial categories subject to effluent guidelines, as well as other industrial categories that could be subject to new guidelines; it has identified 12 categories for further review and selected 3 categories to update or to receive new effluent guidelines.

EPA's Screening Phase Results in a Subset of Industrial Categories for Further Review

EPA's screening phase starts with a review of industrial categories already subject to effluent guidelines—as well as industrial categories that are not—to identify and rank those whose pollutant discharges pose a substantial hazard to human health and the environment.[23] EPA analyzes and ranks industrial categories using pollutant data from facilities in similar industrial classifications. Before it ranks industrial categories in this screening phase, EPA excludes from consideration any industrial categories where guidelines are already undergoing revision or have been revised or developed in the previous 7 years.

For example, EPA announced in its 2010 final effluent guideline program plan that it excluded the steam electric power-generating category from the screening phase because the agency had already begun revising effluent guidelines for this industry.[24] Also in 2010 EPA excluded the concentrated aquatic animal production category (e.g., fish farming) from screening because the agency issued effluent guidelines in 2004.

Steam Electric Power-Generating Industrial Category

The steam electricity power-generating industrial category produces electric power by means of steam generated from fossil fuels (coal, oil, and natural gas) or nuclear fuels. In the process, a given power plant may discharge pollutants such as arsenic and mercury into water bodies. Pollutants from this category persist in the environment and can accumulate in aquatic organisms, wildlife, and humans, with long-term health effects. Effluent guidelines for this industry were first promulgated in 1974 and revised less than a decade later, in 1982.

Source: EPA Documentation.

In ranking industrial categories during the screening phase, EPA considers the extent to which discharged pollutants threaten human health and the environment—the first factor identified in EPA's 2002 draft strategy. EPA compiles information from two EPA sources on the facilities within these industrial categories that discharge wastewater, the pollutants they discharge, and the amount of their discharge: (1) the discharge monitoring report database and (2) the Toxics Release Inventory. [25] From these two sources, EPA estimates the amount and relative toxicity of pollutant discharges from screened industrial categories, converts these estimates into a single "score" of relative toxicity for each industrial category, and uses this score to rank the industrial categories according to the reported hazard they pose.

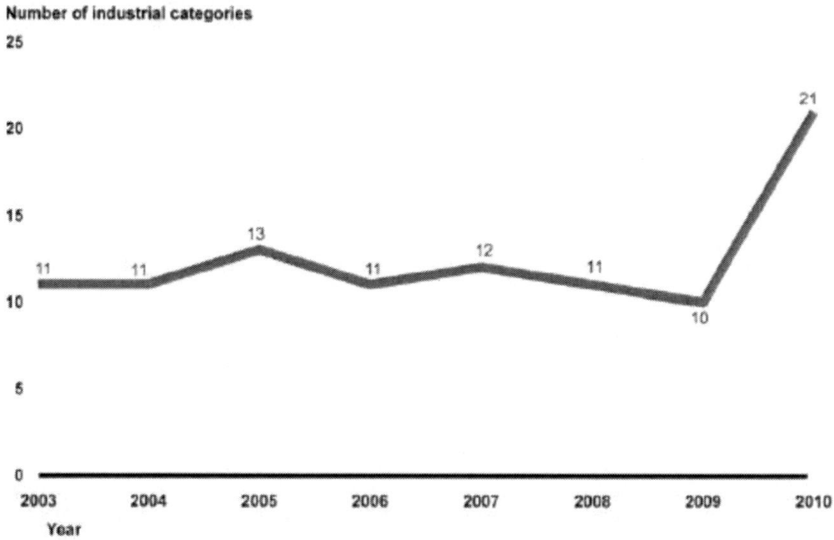

Source: GAO analysis of EPA documents.

Figure 2. Industrial Categories Responsible for 95 Percent of the Total Reported Hazard and Considered for Possible Further Review, 2003-2010.

To determine the relative toxicity of a given pollutant, EPA multiplies the amount (in pounds) of that pollutant by a pollutant-specific weighting factor to derive a "toxic weighted pound equivalent." EPA's ranking of one industrial category relative to other categories can vary depending on the amount of the pollutants it discharges or the toxicity of those pollutants. For example, an industrial category, such as pesticide chemicals, may discharge fewer pounds of pollutants than another category, such as canned and preserved seafood processing, but have a higher hazard ranking because of the relative toxicity of the pollutant chemicals it discharges.

After ranking industrial categories, EPA identifies those responsible for the top 95 percent of the total reported hazard, which is the total of all industrial categories' hazard scores. EPA assigns these industrial categories a high priority for further review in the second phase of its review process. As the relative amounts of their discharges change, the number of industrial categories making up this 95 percent can vary each year with each screening EPA performs. From 2003 through 2009, for example, 10 to 13 industrial categories composed the top 95 percent of reported hazard, whereas in 2010, 21 categories made up the top 95 percent.[26] Figure 2 shows the number of industrial categories that EPA considered for possible further review on the basis of its hazard screening.

After it identifies the industrial categories contributing to 95 percent of reported hazard, EPA takes additional steps to exclude industrial categories before beginning the further review phase. Specifically, the agency may exclude industrial categories on the basis of three criteria:

- *Data used in the ranking process contained errors.* After completing its ranking, EPA verifies the pollutant discharge data from the discharge monitoring reports and Toxics Release Inventory and corrects any errors. For example, according to EPA, the agency has found that facilities have reported the wrong unit of measurement in their discharge monitoring reports, or states have transferred data into the EPA database incorrectly. In such cases, a pollutant discharge may, for example, be reported at a concentration of 10 milligrams per liter but in fact be present at a concentration of 10 micrograms per liter—a thousand-fold lower discharge.
- *Very few facilities account for the relative toxicity of an industrial category.* EPA typically does not consider for further review industries where only a few facilities account for the vast majority of pollutant discharges and the discharges are not representative of the category as a whole. In such cases, EPA states in its effluent guideline program plans that revising individual NPDES permits may be more effective than a nationwide regulation to address the discharge. For example, in 2004, EPA determined that one facility was responsible for the vast majority of discharges of dioxin associated with the inorganic chemicals industrial category. In its effluent guideline program plan for that year, EPA indicated that it would work through the facility's NPDES permit to reduce these discharges as appropriate.
- *Other factors.* EPA considers other factors in addition to those described above to determine if an industrial category warrants further review. According to EPA, one such factor is inadequate data from which to make a clear determination. For example, in its 2010 screening phase, EPA excluded several industrial categories from the further review phase because it did not have conclusive data but said that it would "continue to review" the categories' discharges to determine if they were properly controlled. These industries included pulp, paper, and paperboard; plastic molding and forming; and waste combustors.

Figure 3 illustrates the exclusion process EPA applies in its initial screening phase.

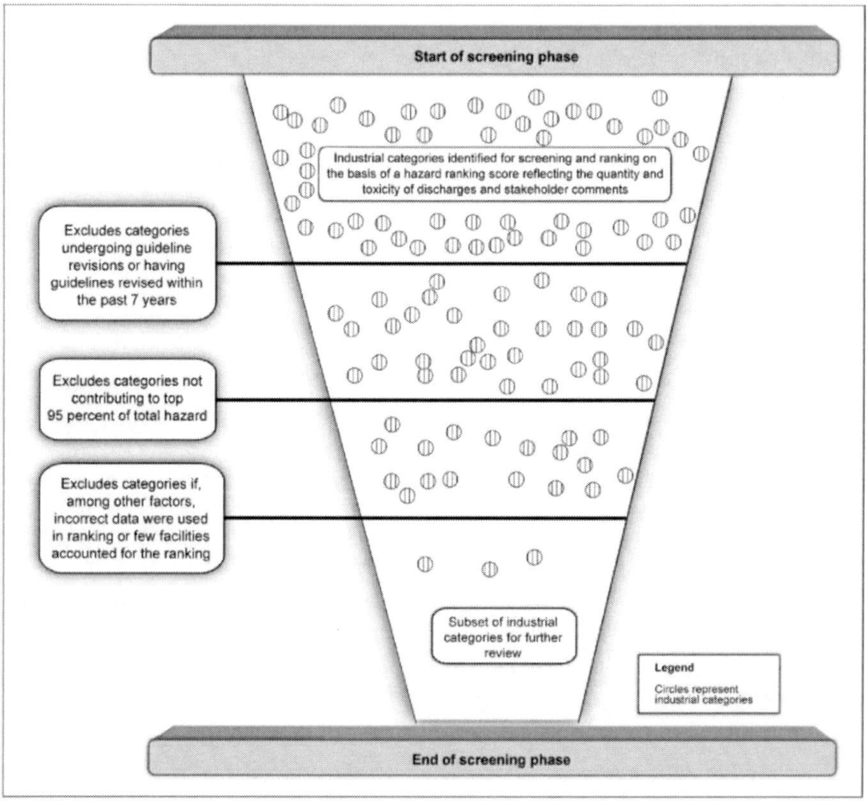

Source: GAO analysis of EPAs screening process.

Figure 3. Criteria Used by EPA during Screening Phase to Exclude Industrial Categories from Further Review.

During the screening phase, EPA uses existing industry classifications as the basis for identifying industrial categories. EPA groups these industry classifications, which are identified by one of two standardized coding schemes, into industrial categories that it then considers for effluent guidelines.[27] If EPA identifies an industrial category that does not have effluent guidelines but has discharges that present a potential hazard, it decides whether the category produces a product or performs a service similar to one subject to existing effluent guidelines. If so, EPA generally considers the former category to be a subcategory of the latter. Conversely, if the products or services differ from categories subject to existing guidelines, EPA considers the category as a potential new category. In either case, EPA may decide that the industrial category warrants further review and, possibly, new effluent guidelines.

Throughout the screening phase, EPA also obtains stakeholder and public input, which may identify industrial categories warranting new or revised effluent guidelines that were not identified by their hazard ranking. Stakeholder and public input comes from EPA's solicitation of comments on its biennial preliminary and final effluent guidelines program plans. For example, in 2004 stakeholders raised concerns about discharges from dental facilities of mercury used in dental fillings; in response, EPA later identified the dental category for further review. On completing the screening phase, the agency lists in its preliminary or final effluent guidelines program plans the industrial categories it has identified for further review. Alternatively, EPA may decide on the basis of its screening criteria that no industrial categories warrant further review.

EPA's Further Review Phase Results in Few Industrial Categories to Consider for Potential New or Revised Effluent Guidelines

In its further review phase, EPA conducts detailed studies of any industrial categories identified in its screening phase, using the four factors listed in its November 2002 draft strategy to determine whether the categories need new or revised effluent guidelines. Since issuing its draft strategy, EPA has selected 12 industrial categories to move beyond the screening phase to the further review phase. Seven of the categories—for example, the pulp, paper, and paperboard category and the petroleum refining category—were identified for further review on the basis of the risk or toxicity of the pollutants they discharge, and 5 were identified for review on the basis of stakeholder concerns. If the categories are already subject to effluent guidelines that EPA set, the agency studies the need to revise effluent limits in the existing guidelines; if the categories are not subject to existing guidelines, EPA studies the need to develop effluent limits and apply them for the first time. Of the 12 categories selected for further review, 8 were already subject to existing effluent guidelines, and 4 were not.

During its further review phase, according to EPA documents, EPA gathers and analyzes more information on the factors identified in its draft strategy. During this phase, EPA typically analyzes information on the hazards posed by discharged pollutants, which corresponds to the first factor in its draft strategy. The data on hazards that EPA obtains and analyzes include: (1) characteristics of wastewater and of facilities; (2) the pollutants responsible for the industrial category's relative toxicity ranking; (3) geographic distribution of facilities in the industry; (4) trends in discharges within the industry, and (5) any relevant economic factors related to the industry.

Airport Deicing Industrial Category

EPA's effluent guidelines for the airport deicing industrial category are the agency's most recent. They apply to the fluids airports use to deice airfield pavement. Discharges from deicing operations may lead to fish kills and reduced species diversity, contaminate drinking water sources (both surface and groundwater), and create noxious odors and discolored water. Airports with 1,000 or more annual jet departures must either control their discharge of ammonia or use non-urea-based airfield pavement deicing fluids. EPA first studied the airport deicing industrial category in the late 1990s to meet the terms of a consent decree. EPA named the category for rule making in 2004, and the agency finalized its regulations in May 2012.

Source: EPA documentation.

During the further review phase, EPA also begins to gather and analyze information on the availability of pollution prevention and treatment technology for the industrial categories reviewed, which corresponds to the second factor identified in its draft 2002 strategy.

Through this analysis, EPA identifies current technologies that industry is using to reduce pollutants, potential new technologies that could be used to reduce pollutants, or both.

Table 3 summarizes EPA's consideration of treatment technologies for the 12 industrial categories that proceeded to the further review phase. For example, EPA studied one technology used by the ore mining and dressing industrial category and several current technologies for the coalbed methane category.

Table 3. Consideration of Treatment Technology during Further Review and Resulting Key Agency Decisions

Industrial category	Period for further review[a]	Consideration of treatment technology during further review	Key agency decisions after further review
Categories with existing effluent guidelines			
Chlorine and chlorinated hydrocarbons manufacturing[b]	2005-present	EPA did not prepare a written study for this industrial category. Instead it used site visits and sampled wastewater to collect data on pollutant quantities and conducted site visits to design an industry-led voluntary sampling plan. EPA officials stated that through implementation of the sampling plan, the agency discovered that most participating facilities used treatment technology that was effectively controlling their pollutant discharges	According to EPA officials, the agency does not plan to initiate rule making for this industrial category because only one facility is responsible for the majority of the pollutant discharge.
Coalbed methane extraction[c]	2005-2010	In its 2010 study, EPA presented an overview of seven treatment methods used by this.	On the basis of the study's findings, EPA in 2010 announced the start of rule making

Table 3. (Continued)

Industrial category	Period for further review[a]	Consideration of treatment technology during further review	Key agency decisions after further review
		industry, depending on the type of wastewater pollutant produced during the extraction process	for effluent guidelines for this category. The agency plans to propose the rule in 2013.
Coal mining	2006-2008	In its 2008 study, EPA described two treatment technologies associated with the most common pollutant discharges resulting from coal mining processes.	On the basis of the study's findings, EPA decided to take no further action on this category, stating that the existing guidelines were appropriate to address the industry's discharges
Ore mining and dressing	2008-2011	EPA prepared a study published in 2011, which included a review of one currently used treatment technology: high density sludge recycling. The review included an overview of this treatment, where it was used in the United States, and permit requirements for facilities that used the technology.	EPA in 2011 decided to take no further action on ore mining and dressing. The agency found that a small number of facilities were responsible for the majority of discharges and decided to address the pollutants the NPDES permit process with permitting or through compliance and enforcement activities.
Organic chemicals, plastics, and synthetic fibers	2003-2004	EPA's study of this category included general descriptions of treatments currently used by the industry, as well as a more detailed discussion of pollution prevention	EPA in 2004 decided to take no further action for the category because of the small number of facilities discharging significant amounts of toxic pollutants.

Industrial category	Period for further review[a]	Consideration of treatment technology during further review	Key agency decisions after further review
		and wastewater treatment technologies used to minimize the amount of dioxin in wastewater streams.	
Petroleum refining	2003-2004	In its study of the petroleum refining category, EPA presented general information on current treatments used by refineries to treat wastewater produced and on additional pollution control alternatives.	EPA in 2004 decided to take no further action on petroleum refining because it found that most petroleum refining facilities were not discharging toxic pollutants. For the few facilities that were, the agency said it would seek changes through the NPDES permit process.
Pulp, paper, and paperboard	2005-2006	In its 2006 study, EPA identified technologies used in a laboratory setting, pilot programs, and industry to remove metals in wastewater from pulp and paper mills.	On the basis of findings from the further review phase, including that some available technologies were site-specific and not readily adaptable industrywide, EPA in 2006 decided to take no further action on this category.
Steam electric power generating	2005–2009	As part of its 2009 study, EPA focused primarily on technologies associated with two sources (coal-ash-handling operations and wastewater	On the basis of findings from its study, EPA in 2009 decided to begin revising the effluent guidelines for this category. Proposed revisions to existing
		produced from specific air pollution control systems) because these sources	guidelines are expected in November 2012, with final action expected by April

Table 3. (Continued)

Industrial category	Period for further review[a]	Consideration of treatment technology during further review	Key agency decisions after further review
		account for a significant amount of the pollutants discharged by the industry.	2014.
New industrial categories considered			
Dental facilities	2006-2008	EPA issued a study of dental facilities in 2008. The study discussed best management practices for dental facilities to reduce their discharge of dental amalgam containing mercury. In particular, the study reviewed facilities' continuing use of amalgam separators.	EPA In 2008 announced that it would not pursue a rule making and would instead work with stakeholders, including the American Dental Association and state water agencies, on a voluntary discharge reduction program. In 2010, however, the agency reversed its decision after assessing the progress made under the voluntary reduction program. EPA expects to propose effluent guidelines in 2012.
Drinking water treatment	2004-2011	EPA reviewed available treatment technologies as part of a 2011 study of the industry. Because drinking water treatment operations vary—in types of contaminants found at different plants, for example—the study presented a range of	EPA took no further action on this category because the agency found that discharges from the category could best be addressed by adding limits to specific NPDES permits. In addition, according to agency officials, the 2011 technical report

Industrial category	Period for further review[a]	Consideration of treatment technology during further review	Key agency decisions after further review
		treatment approaches.	can provide information on technologies for state permit writers for drinking ater treatment facilities.
Pharmaceuticals management[d]	2006-present	According to EPA officials, a study of the industrial category is still under way. Officials said that because of the nature of the industry, they are pursuing a "front-end" strategy to prevent flushing of unused pharmaceuticals into wastewater systems. Given this approach, treatment technologies will not be a primary focus of the agency's further review process.	EPA continues to work on its study, although according to agency officials, no further action is expected toward developing new effluent guidelines for this industry. Instead, the agency plans to update draft guidance issued in August 2010 on best management practices for unused pharmaceuticals at health care facilities.
Tobacco products processing	2004-2006	In 2006, EPA issued a study on this industry, which provided an overview of the treatment process typically used by tobacco products facilities.	The agency found that the category comprises primarily indirect dischargers; that pollutant loads are low; and, according to EPA officials, wastewater treatment plants remove 96 percent of the loads. EPA in 2006 therefore decided to take no further action

Source: GAO analysis of EPA documents.

[a] The dates of further review (1) start with the year when EPA announced in an effluent guidelines program plan that it would conduct a study and (2) end with the year when EPA completed that work.

[b] Chlorine and chlorinated hydrocarbons manufacturing is a subcategory of two existing effluent guidelines categories: (1) organic chemicals, plastics, and synthetic fibers and (2) inorganic chemicals.

[c] Coalbed methane extraction is not covered by an existing effluent guidelines category, although EPA considers the industry a new subcategory of the oil and gas extraction category

[d] Pharmaceuticals management is not to be confused with pharmaceutical manufacturing.

During its further review phase, EPA also obtains and analyzes information related to the cost, affordability, and performance of technologies, the third factor in its strategy. To do so, EPA examines the cost and performance of applicable technologies, changes in production processes, or prevention alternatives that may reduce pollutants in the industrial category's discharge. As part of its cost analysis, the agency considers the affordability or economic achievability of any identified technologies, production processes, or prevention alternatives. To assess the performance of technologies, EPA considers the results of the treatment technologies used in tests or actual operations—information the agency obtains from published research papers and internal and external sources, including site visits and surveys of industrial facilities.[28] During its further review of the steam electricity power-generating industry, for example, EPA sampled wastewater directly at power plants, surveyed plant operators about which technologies they were using to minimize pollutant discharges and at what cost, and sought information on other potential treatment technologies.

At the conclusion of its further review of an industrial category, EPA decides whether it is feasible and appropriate to revise or develop effluent guidelines for the category, a decision that includes gathering information on whether an effluent guideline is the most efficient and effective approach to manage the discharges, the fourth factor in EPA's draft strategy. As shown in table 3, for example, EPA decided that the drinking water treatment industrial category did not require effluent guidelines but that the agency's study could act as a resource for state permit writers as they issue permits for drinking water facilities. Or, as also shown in table 3 for coalbed methane, EPA decided to develop guidelines that it plans to propose in 2013. Some of the information EPA can consider during this decision making, and some of the information related to the fourth factor in its strategy, is the extent to which existing effluent guidelines could be revised to eliminate inefficiencies or impediments to technological innovation or to promote innovative approaches. Specifically, EPA considers whether another way exists—either regulatory or voluntary—to decrease pollutant discharges. For example, after the further review of the dental facility category in 2008, EPA decided not to develop effluent guidelines but to instead work with the American Dental Association and state water agencies on a voluntary reduction program to reduce pollutant discharges from dental facilities. It later changed its decision because the voluntary effort was shown to be ineffective, and the agency plans to issue effluent guidelines in 2012.

Coalbed Methane Extraction Industrial Category

Coalbed methane extraction requires the removal of groundwater to facilitate the flow of natural gas to the surface, unlike extraction of conventional natural gas. The industrial category is currently unregulated. Through a nationwide survey, EPA determined that the 750 coalbed methane extraction projects across the country use 56,000 individual wells. Total dissolved solids in the discharge from coalbed methane extraction wells are EPA's chief pollutant of concern. The photograph shows a tanker truck loading groundwater collected during methane extraction. According to agency documents, EPA plans to propose an effluent guideline for this portion of the oil and gas extraction industrial category in 2013.

Source: EPA documentation.

It takes EPA, on average, 3 to 4 years to complete the further review phase for an industrial category.

As of July 2012, EPA had identified three industrial categories for which it had decided to revise effluent guidelines—steam electric power generating—or to develop new effluent guidelines—coalbed methane extraction and dental facilities.[29] According to agency documents and officials,

EPA has chosen to take no action on the other 9 of the 12 categories it has further reviewed since 2002.

Focus on Limited Hazard Data to the Exclusion of Technology Information May Have Led EPA to Overlook Industrial Categories for Pollution Reduction

Limitations in the screening phase of EPA's review process may have caused the agency to overlook some industrial categories that warrant new or revised effluent guidelines and thus hinder the effectiveness of the effluent guidelines program in advancing the goals of the Clean Water Act.

First, the data EPA uses in the screening phase has limitations that may cause the agency to omit industrial categories from further review or regulation. Second, EPA has chosen to focus its screening phase on the hazards associated with industrial categories, without considering the availability of treatment technologies or production changes that could reduce those hazards.

The screening phase of the process may thus exclude some industrial categories for which treatment technologies or production changes may be available to serve as the basis for new or revised effluent guidelines.

Limitations in Hazard Data May Have Caused EPA to Overlook Industrial Categories

The two sources EPA relies on during its initial screening process—discharge monitoring reports and the Toxic Release Inventory—have limitations that may affect the agency's ability to accurately rank industrial categories for further review on the basis of the human health and environmental hazards associated with those categories.

Data from industrial facilities' discharge monitoring reports have the benefit of being national in scope, according to EPA documents, but according to agency officials and some experts we spoke with, these data have several limitations that could lead the agency to underestimate the hazard caused by particular industries.

Specifically:

- *The reports contain data only for those pollutants that facilities' permits require them to monitor.* Under NPDES, states and EPAoffices issue permits containing limits for pollutant discharges, but those permits may not include limits for all the pollutants that may be discharged, as for example, if those pollutants are not included in the relevant effluent guidelines or need not be limited for the facility to meet state water quality standards.[30] If a pollutant is not identified in a permit, and hence not reported on discharge monitoring reports, it would not be part of EPA's calculation of hazard and would not count toward the ranking of industrial categories.
- *The reports do not include data from all permitted facilities.* Specifically, EPA does not require the states to report monitoring results from direct dischargers classified as minor. According to EPA, the agency in 2010 analyzed data for approximately 15,000 minor facilities, or about 37 percent of the 40,500 minor facilities covered by NPDES permits. As a result, the pollutants discharged by the remaining 25,500 minor dischargers would not be counted as part of the relative toxicity rating and could contribute to undercounting of pollutants from those industrial categories. For example, most coal mining companies in Pennsylvania and West Virginia are considered minor dischargers whose pollutants would not count toward the ranking of that industrial category.
- *The reports include very limited data characterizing indirect discharges from industrial facilities to wastewater treatment plants,* according to EPA documents. Thus, the data do not fully document pollutants that, if not removed by a wastewater treatment plant, are discharged. These data are not incorporated into EPA's calculations of hazard for each industrial category, and thus result in underestimated hazards.[31]

EPA documents and some experts we contacted also stated that data collected in the Toxics Release Inventory are useful to identify toxic discharges. Nevertheless, according to the agency and experts, these inventory data have limitations that may cause EPA to either overestimate or underestimate the relative toxicity of particular industrial categories. The limitations they identified include the following:

- *The data reported are sometimes estimates and not actual monitored data.* In some cases, the use of an estimate may overreport actual pollutant discharges. For example, some industry experts said that to be conservative and avoid possible liability, some facilities engaging in processes that produce particularly toxic pollutants, such as dioxin, may report the discharge of a small amount on the basis of an EPA-prescribed method for estimating such discharges even if the pollutant had not been actually monitored.

- *Not all facilities are required to report to the inventory*, which may lead to undercounting the discharges for the industrial categories of which the facilities are a part. Facilities with fewer than 10 employees are not required to report to the inventory, and neither are facilities that do not manufacture, import, process, or use more than a threshold amount of listed chemicals. For example, facilities that manufacture or process lead or dioxin do not need to report to the inventory unless the amount of chemical manufactured or processed reaches
- 10 pounds for lead or 0.1 grams for dioxin.

Despite the limitations of these data sources, EPA officials said that discharge monitoring reports and the Toxic Inventory Release are the best available data on a national level.

Experts we interviewed also generally supported the continued use of these data sources despite their limitations.

An EPA official responsible for the screening and review process said that EPA could not quantify the effect of the missing data on its ranking and setting of priorities for industries without time-consuming and expensive collecting of data directly from industrial facilities. Still, agency officials agreed that the data limitations can lead to under- or overestimating the hazard of discharges from industrial categories, which could in turn affect the rankings of these categories and potentially result in different categories advancing for further review and potential regulation.

EPA's Screening Phase Does Not Consider Treatment Technologies, Omitting Some Industrial Categories from Further Review

EPA's primary focus during its screening phase is the relative hazard posed by industrial categories, without consideration of available treatment technologies that could be used as the basis for revised effluent guidelines to help reduce pollutant discharges. Because EPA sets the cutoff point in its screening process as industrial categories contributing to 95 percent of total reported hazard, the agency does not consider for further review the categories contributing to 5 percent of the total reported hazard. Although this percentage is low, the categories involved constitute the majority of all industrial categories with effluent guidelines. EPA does not conduct a further review for these and other industrial categories that it has excluded for other reasons, meaning that EPA does not examine them for the availability of more-effective treatment technologies. As previously noted, the Ninth Circuit Court held in 2008 that EPA does not have a mandatory duty to consider technology in its screening process but stated that the act strongly suggests that any review to determine whether revision of effluent guidelines is appropriate should contemplate technology-based factors. Regardless of whether EPA is required to do so, the agency is not considering technology for these industrial categories, and hence EPA cannot ensure that the facilities in these categories are using the best available treatment technology.

According to our analysis of EPA's planning documents for the effluent guidelines program, since the agency in 2003 began using its current screening process, more than half the industrial categories with effluent guidelines did not advance beyond the screening phase in any year from 2003 to 2010. The reason these categories did not advance was that, during a given 2-year screening cycle, the relative toxicity of their pollutant discharges did not put them among the top 95 percent of dischargers. As a result, these industrial categories were excluded from further review before EPA would have analyzed the availability of more-advanced treatment technologies or production processes. As figure 4 shows, from 2003 through 2010, out of the 57 industrial categories with existing effluent guidelines at the time of review, EPA excluded 35 in each of the four biennial screening cycles, thus omitting them from further review of the availability of treatment technologies or production processes that could reduce hazards from discharges.32 (See app. III for further information on the industrial categories that have and have not come within the top 95 percent since 2003.)

As noted in 4 of the 17 interviews with experts we interviewed from academia, industry, nonprofit organizations, and state and local water quality agencies, consideration of treatment technologies is especially important for older effluent guidelines because either the industrial categories or the treatment technologies would have been more likely to change, making it possible that new, more advanced treatment technologies are available. As table 4 shows, however, effluent guidelines have not been revised since the 1980s for 8 of the 35 industrial categories that have been excluded from further review. Further, 25 of the 35 effluent guidelines for categories that were excluded from further review have not been revised since 1995 or earlier. Battery manufacturing, for example, has not been through the further review phase since EPA began using its current screening and review process. Yet according to state officials we interviewed, the effluent guidelines for this category apply to older battery types and do not address wastewater from the manufacture of newer battery types, such as those made with lithium. In addition, even in cases where EPA has revised the effluent guideline for an industrial category, the revision may have addressed just a portion of the guideline. For example, EPA may add pollutants or change the limits for a particular industrial category or add a new subcategory. Thus, the guidelines for the category as a whole may not have been revised since the guidelines were originally promulgated. Table 4 shows the 35 industrial categories that were excluded from further review over the last 8 years, the year effluent guidelines were promulgated, and the year the categories' guidelines were most recently revised.

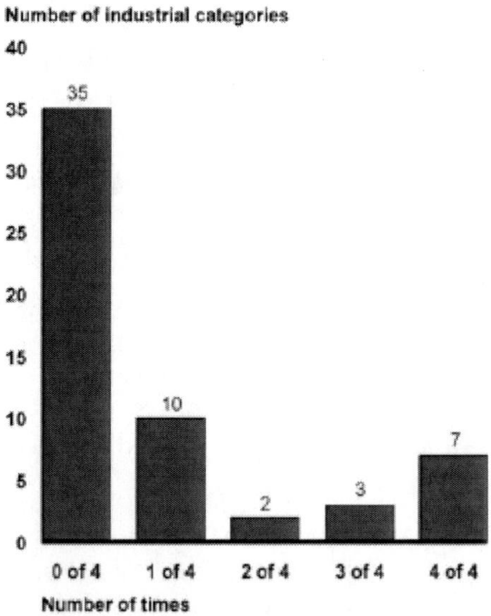

Source: GAO analysis of EPA data.

Note: The total number of industrial categories with effluent guidelines during this period was 57.

Figure 4. Number of Times Existing Industrial Categories Were in the Top 95 Percent of Total Reported Hazards in the Four Biennial Screening Phases from 2003 through 2010.

Table 4. Regulated Industrial Categories Excluded in the Screening Phase from Further Review, 2003-2010

Industrial category	Year promulgated	Year most recently revised	Number of years elapsed since mostrecent revision
Porcelain enameling	1982	1985	27
Electrical and electronic components	1983	1985	27
Electroplating	1981	1986	26
Copper forming	1983	1986	26
Metal finishing	1983	1986	26
Battery manufacturing	1984	1986	26
Aluminum forming	1983	1988	24
Nonferrous metals forming and metal powders	1985	1989	23
Asbestos manufacturing	1974	1995	17
Canned and preserved fruits and vegetables processing	1974	1995	17
Canned and preserved seafood processing	1974	1995	17
Dairy products processing	1974	1995	17
Ferroalloy manufacturing	1974	1995	17
Glass manufacturing	1974	1995	17
Grain mills	1974	1995	17
Soap and detergent manufacturing	1974	1995	17
Sugar processing	1974	1995	17
Ink formulating	1975	1995	17
Paint formulating	1975	1995	17
Paving and roofing materials (tars and asphalt)	1975	1995	17
Explosives manufacturing	1976	1995	17
Gum and wood chemicals	1976	1995	17
Hospital	1976	1995	17
Carbon black manufacturing	1978	1995	17
Leather tanning and finishing	1982	1996	16
Pharmaceutical manufacturing	1983	2003	9
ronand steel manufacturing	1982	2005	7
Transportation equipment cleaning	2000	2005	7
Coil coating	1982	2007	5
Concentrated animal feeding operations	2003	2008	4
Photographic	1976	No revisions	No revisions
Metal products and machinery	2003	No revisions	No revisions
Concentrated aquatic animal production	2004	No revisions	No revisions

Table 4. (Continued)

Industrial category	Year promulgated	Year most recently revised	Number of years elapsed since mostrecent revision
Meat and poultry products	2004	No revisions	No revisions
Construction and development	2009	No revisions	No revisions

Source: GAO analysis of EPA documentation.

Note: In a given year, EPA may have revised just a portion of the effluent guideline for an industrial category. For example, EPA may have added pollutants or changed the pollutant limits for a particular industrial category or added a new subcategory.

Our survey of state water quality directors, who are responsible for NPDES permits, also identified industrial categories that have been omitted from EPA's further review phase even when treatment technologies may be available. Specifically, state officials identified nine industrial categories that they think pose significant risk and have treatment technologies or pollution prevention practices available to mitigate that risk, categories for which the effluent guidelines should be revised. Further, state officials generally thought that industries could implement the technologies without financial hardship. Nevertheless, EPA has excluded these industrial categories from further review because they did not contribute to the top 95 percent of total reported hazard. At least one state director identified one or more of the following nine industrial categories as needing revised effluent guidelines, noting that their pollutants were hazardous to human health or the environment and technologies were available to further reduce these hazards: canned and preserved seafood processing; dairy products processing; electrical and electronic components; electroplating; grain mills; meat and poultry products; metal finishing; pharmaceutical manufacturing; and sugar processing. With regard to metal finishing, for example, state officials said that existing guidelines reflect processes no longer in use and do not address newer and more common production techniques and associated pollutants. In contrast, state officials agreed with EPA's efforts to revise or develop new effluent guidelines for certain other industrial categories, including steam electric power generation and airport deicing. (For more information on our survey and its results, see appendix II.)

In its 2002 draft strategy, EPA recognized the importance of including treatment technology in its screening phase but later stated that it was unable to develop an approach it deemed feasible for gathering such information. The draft strategy included treatment technology as one of the factors that EPA would use to screen industrial categories to determine if they needed new or revised effluent guidelines. According to the draft strategy, EPA was to obtain information on available treatment technologies and pollution prevention practices by regularly reviewing trade publications; participating in professional conferences; and consulting with permit writers, industry representatives, and the public. EPA initially pursued this approach, but in 2003 concluded that gathering the data needed to perform a meaningful screening-level analysis for technology was much more resource intensive than anticipated[33] and restricted the screening phase to comparing the degree of hazard posed by various industrial categories.[34] Yet without treatment technology data, the agency cannot be confident that the effluent guidelines program is meeting the Clean Water Act's goal of applying the best available technologies economically achievable or that the program reflects advances in the technologies used to reduce pollutants in wastewater.

EPA IS ADDING HAZARD DATA SOURCES BUT IS NOT FULLY USING POTENTIAL SOURCES OF INFORMATION ON TREATMENT TECHNOLOGIES

EPA has begun to take actions to improve the hazard data it uses in its screening of industrial categories, but it is not fully using potential sources of information on treatment technologies for consideration in this screening. According to program officials, EPA has recognized that its screening phase has resulted in the same industries rising repeatedly to the top of its hazard rankings. Program officials said that they are considering changes to their screening approach to identify additional industrial categories for further review. The primary change, the officials told us, would be to rank categories according to toxicity every 2 years, rather than annually, and to supplement that ranking with a targeted analysis of additional sources of data. To develop such revisions, officials from EPA's effluent guidelines program engaged in an informal "brainstorming" exercise within the agency and identified several sources of data on new and emerging pollutants, sources that officials think could help target industrial categories for further review. EPA officials said they will propose revisions to the review process in the 2012 preliminary effluent guidelines program plan they expect to issue late in 2012.

To mitigate the limitations with hazard data that EPA currently experiences, the agency has taken several steps to obtain new sources of information and to improve existing sources. Using additional sources of data is consistent with suggestions made to us by several academic and governmental experts we interviewed that other sources of hazard data may be useful to the agency, including additional monitoring data and data on the quality of water bodies receiving wastewater discharges.[35] The new data sources would broaden the hazard data considered in the screening phase. Among the sources EPA intends to pursue for future use are the following:

- a 2009 EPA survey of sludge produced by wastewater treatment plants to identify pollutants entering these plants, indicating that they are not being treated by an industrial facility and might need regulation;
- a review of action plans prepared under EPA's Office of Pollution Prevention and Toxic Substances for specific chemicals of emerging concern to identify pollutants that are likely to be discharged to waters by industrial point sources;
- a review of all EPA air pollution regulations issued within the last 10 to15 years to identify new treatment processes that could add to or change the pollutants in wastewater streams;[36] and
- a review of data and information available concerning industries that EPA is considering for a proposed expansion of required reporting for the Toxics Release Inventory.

EPA is also drafting a rule that would increase the information EPA receives electronically from discharge monitoring reports from NPDES permittees and permitting authorities. According to officials with the effluent guidelines program, increased electronic reporting would result in a more complete and accurate database and improve their access to the hazard data from facilities' discharge monitoring reports, thereby improving the screening of industrial categories. For example, according to EPA officials, data on minor facilities that are not currently reported into the discharge monitoring database used in the screening process would be reported under the electronic reporting rule, as sent to the Office of Management and Budget for review.[37]

EPA recognizes the need to use information on treatment technologies in the screening phase to improve its process and has taken some initial steps to develop a database of such information, but it has not made full use of potential data sources. EPA started to gather information on treatment technology in 2011, contracting with consultants to obtain relevant literature for the database. In its comments on a draft of this report, the agency said that it will expand on this work in 2013 and 2014 once new fiscal year operating plans are in place. According to agency officials, a thorough analysis of the literature would give the program an updated technology database, which would help in identifying advances in technologies in use or with potential use in industrial categories, which, on the basis of these advances, may in turn warrant further review. They noted that in the 1980s and 1990s, the program used such information from an agency database but that the database had become outdated.

In more than half of our interviews (10 of 17), experts told us that EPA should consider technology in its screening phase,[38] and some of them suggested the following two approaches for obtaining this information:

- *Stakeholder outreach.* Experts suggested that key stakeholders could provide information on technology earlier in the screening process. Currently, EPA solicits views and information from stakeholders during public comment periods following issuance of preliminary and final effluent guidelines plans. According to experts, EPA could obtain up-to-date information and data from stakeholders beyond these formal comment periods. For example, EPA officials could (1) attend annual workshops and conferences hosted by industries and associations, such as engineering associations, or host their own expert panels to learn about new treatment technologies and (2) work with industrial research and development institutes to learn about efforts to reduce wastewater pollution through production changes or treatment technologies.
- *NPDES permits and related documentation.* Experts suggested that to find more information on treatment technologies available for specific pollutants, EPA could make better use of information in NPDES permit documentation. For example, when applying for NPDES permits, facilities must describe which pollutants they will be discharging and what treatment processes they will use to mitigate these discharges. Such information could help EPA officials administering the effluent guidelines program as they seek technologies to reduce pollutants in similar wastewater streams from similar industrial processes. Similarly, information from issued NPDES permits containing the more stringent water quality-based limits— which may lead a facility to apply more advanced treatment

technologies—could suggest the potential for improved reductions. Further, information in fact sheets prepared by the permitting authority could also furnish information on pollutants or technologies that could help EPA identify new technologies for use in effluent guidelines.

According to EPA officials, these two sources of information have not been extensively used. They said that they would like to obtain more stakeholder input during screening and review, but they have limited time, resources, and ability to work with stakeholders. They noted that the effluent guidelines program does assign staff members responsibility for keeping up with technologies and developments in specific industrial categories. They also said that the NPDES information suggested by experts is not current or readily available for use by the program.

Our analysis of NPDES information, however, showed that EPA has not taken steps to make the information available for use by the effluent guidelines program. For example, the standard list of treatment processes on the NPDES application form has not been updated since 1980, and EPA officials said it was out of date. Yet EPA has not updated this information or provided it to the effluent guidelines program for use in screening available technologies. EPA could have done so through a second rulemaking effort under way to improve NPDES data—in which EPA is updating NPDES application forms to make them more consistent with NPDES regulations and current program practices—but chose not to. Agency documents about this rulemaking described it as modifying or repealing reporting requirements that have become obsolete or outdated over the past 20 years and modifying permit documentation procedures to improve the quality and transparency of permit development.

Nonetheless, effluent guidelines program officials said that they did not request potential NPDES permit updates relevant to their program because the scope of this rulemaking was too narrow. EPA's Office of Wastewater Management, which is responsible for the rulemaking, confirmed that the scope of the proposed rule is to be narrow and not call for states or permittees to provide new information.

Further, fact sheets or similar documentation that NPDES permit writers develop describing the basis for permit conditions are not stored in EPA's electronic NPDES database and are therefore difficult to obtain and analyze, according to program officials. Instead, these NPDES documents are now maintained by the authorized states or EPA regions and are not readily accessible to the effluent guidelines program.

Program officials said that electronic transmission of fact sheets or information about the basis for permit limits could be useful in identifying treatment technologies, although the scope of the electronic reporting rulemaking did not include such documents or information. Officials from the Office of Enforcement and Compliance Assurance, the office responsible for this rulemaking, told us that they discovered such wide variability among the states' practices for gathering and managing NPDES information like fact sheets or the basis for permit limits that it would be difficult to call for electronic reporting of such information.

CONCLUSION

EPA and the nation have made great strides in reducing the pollutants in wastewater discharged from point sources, such as industrial facilities, since the Clean Water Act was passed. EPA's effluent guidelines program has been key in contributing to these results by establishing national uniform limits on pollutant discharges for various industrial categories. Progress within the program has slowed, however, and numerous effluent guidelines for particular industrial categories have not been revised for 2 or 3 decades, although the act calls for EPA to routinely review its effluent guidelines and update or add to them as appropriate. EPA's approach for screening and further reviewing industrial categories, as currently implemented, has not identified many categories for the agency to consider for new or revised guidelines, and the screening process has identified many of the same industrial categories year after year. EPA's approach focuses its resources on the most hazardous sources of pollution, but its reliance on incomplete hazard data during the screening phase has limited the results of the approach, as has EPA's inability to thoroughly collect treatment technology data within its resource constraints. Under EPA's current approach, most industrial categories have not received a detailed further review examining the availability of more-effective treatment technologies. According to some experts, consideration of treatment technologies is especially important for older effluent guidelines because changes in either the industrial categories or the treatment technologies are more likely to have occurred, making it possible that new, more advanced and cost-effective treatment technologies have become available. EPA has recently taken steps to obtain more information on treatment technologies for use in its screening phase—which could help make up for limitations in the hazard data it currently uses—but it has not taken steps to improve and gain access to technology information from the NPDES program. Further, EPA is reconsidering its approach to its screening and review process—initially documented in its draft strategy that was never finalized—but has not analyzed a range of possible sources of data to improve the program, including taking full advantage of the NPDES database, obtaining relevant stakeholder input, and reviewing older effluent guidelines for changes in either the industry or available treatment technologies. Without evaluating a range of new sources of relevant information, officials with the effluent guidelines program cannot ensure that the reconsidered approach can be implemented or that it optimizes the agency's ability to consider technology in the screening process. Most important, without a more thorough and integrated screening approach that both improves hazard information and considers treatment technology data, EPA cannot be certain that the effluent guidelines program is reflecting advances in the treatment technologies used to reduce pollutants in wastewater.

RECOMMENDATIONS FOR EXECUTIVE ACTION

To improve the effectiveness of EPA's efforts to update or develop new effluent guidelines, we recommend that the Administrator of EPA direct the effluent guidelines program to take the following three actions, as it considers revisions to its screening and review process:

- Identify and evaluate additional sources of data on the hazards posed by the discharges from industrial categories.
- Identify and evaluate sources of information to improve the agency's assessment in the screening phase of treatment technologies that are in use or available for use by industrial categories, including better use of NPDES data.
- Modify the screening phase of its review process to include thorough consideration of information on the treatment technologies available to industrial categories.

AGENCY COMMENTS AND OUR EVALUATION

We provided a draft of this report to EPA for review and comment. In its written comments, EPA said that our report adequately describes the agency's effluent guidelines program and agreed in principle with two of the report's recommendations but disagreed with the third recommendation. EPA also provided several technical comments, which we have incorporated as appropriate.

Regarding our first recommendation, that EPA identify and evaluate additional sources of data on the hazards posed by industrial discharges and factor these into its annual reviews, EPA agreed that additional sources of such data are valuable. For this reason, EPA said, it began collecting new sources of hazard information in 2011, which the agency is using in its 2012 annual review. EPA also said that its preliminary 2012 effluent guideline program plan will solicit additional ideas for new hazard data sources from the public and industry stakeholders. We described EPA's ongoing and planned efforts in our report, but because the agency has not yet published its preliminary 2012 effluent guideline program plan, we cannot determine the extent to which these efforts address the limitations we identified in its hazard data. Likewise, we are not able at this time to confirm that EPA will solicit additional sources of such data from stakeholders. We support EPA's stated intent to identify and evaluate additional sources of hazard data and retain our recommendation, reinforcing the need for the agency to continue the efforts it has begun.

Regarding our second recommendation, that EPA should identify and evaluate additional sources of information to improve its assessment of treatment technologies for industrial dischargers, EPA agreed that treatment technology information is useful to its program. The agency added that, given the importance of new treatment technology information, in 2011 it initiated efforts to gather more treatment information across all industry categories and will be expanding on this work in 2013 and 2014, once new fiscal year operating plans are in place. We described EPA's initiative to obtain and review technical literature on treatment technology in our report. We nevertheless believe that EPA could use other sources of information on treatment technology, including information associated with NPDES permits, as described in the report. We continue to believe that EPA should identify and evaluate these and other sources of information on treatment technologies, with the goal of ensuring that the agency's effluent guidelines reflect the best available treatment technologies that are economically achievable.

Regarding our third recommendation, that EPA modify the screening phase of its review process to include a thorough consideration of information on the treatment technologies available to industrial categories, EPA agreed that factoring treatment technology information into its reviews is valuable. The agency said, however, that the recommendation was not workable in the context of the agency's current screening phase, noting that such an effort would be very resource intensive. Our concern is that EPA's current screening phase, while targeted toward high-risk industries, does not ensure that effluent guidelines incorporate the best available treatment technologies that are economically achievable.

We acknowledge that evaluating technologies for all existing industrial categories could be difficult for EPA to accomplish on an annual basis under its current approach.

Our recommendation, however, did not specify that such an evaluation be done every year. For example, EPA could commit to a detailed study of the technologies in use and available to an industrial category on a periodic basis (i.e., every 5-10 years). As noted in our report, EPA's 2002 draft strategy recognized the importance of evaluating treatment technologies in its screening phase, and the Court of Appeals for the Ninth Circuit held that, while not mandatory, the Clean Water Act strongly suggests that in determining whether the revision of effluent guidelines is appropriate—which begins with the screening phase—the agency should contemplate technology-based factors.

However, we are not aware of any detailed EPA evaluation of options for considering technology during the screening phase since the agency announced in 2003 that performing a meaningful screening-level analysis of the availability of treatment technologies as planned in the draft strategy was "much more difficult than anticipated."

We believe that, nearly a decade later, EPA should, within the constraints of available resources, evaluate current options to consider such technologies in its screening phase.

Furthermore, given its efforts to develop and update its technology information, we believe that EPA should clarify how it plans to incorporate this information in its screening phase.

David C. Trimble
Director,
Natural Resources and Environment

APPENDIX I: SCOPE AND METHODOLOGY

To examine the process the Environmental Protection Agency (EPA) follows to screen and review industrial categories and the results of that process, we reviewed the Clean Water Act and relevant court decisions and agency documents, interviewed agency officials and experts, and documented the steps EPA has taken to screen particular industrial categories for possible new or revised effluent guidelines. Specifically, we reviewed relevant portions of the Clean Water Act to determine EPA's responsibilities regarding the effluent guidelines and pretreatment programs. We analyzed several court decisions that ruled on challenges to EPA's effluent guidelines program to determine what, if any, impact they had on the agency's screening and review process. Further, we interviewed officials in EPA's Engineering and Analysis Division to learn how the agency has used the process to screen and review industries. We focused our review on the results of the process EPA used from 2003 through 2010 in order to examine the approach it developed after the publication in November 2002 of its draft *Strategy for National Clean Water Industrial Regulations: Effluent Limitation Guidelines, Pretreatment Standards, and New Source Performance Standards*. By the end of our review, EPA had not yet published a preliminary or final effluent guideline program plan for the 2011-2012 planning cycle.

To document the results of EPA's process, we examined the agency's screening decisions for all industrial categories from 2003 through 2010. Specifically, we examined EPA's final effluent guideline plans and technical support documents for 2004, 2006, 2008, and 2010 and the agency's website to identify screening decisions and subsequent studies associated with particular industries. We examined these studies to identify those industries that EPA subjected to further review, which included an examination of available treatment technologies. Specifically, we examined preliminary and detailed studies for the 12 industries that EPA advanced beyond the screening phase into further review and selected 7 of them for more robust analysis to document how EPA had applied the process to those industries. The 7 industries were ore mining and dressing, coalbed methane extraction, steam electric power generation, chlorine and chlorinated hydrocarbon, drinking water treatment, pharmaceuticals management, and dental facilities. That analysis included in-depth interviews with EPA staff assigned to those industrial categories. These 7 industrial categories met our selection criteria that they be active or recently active, that is, that EPA was reviewing them or had made a decision to proceed or not to proceed with a rulemaking as recently as 2011 or 2012. We also documented the current status of any regulatory actions or other steps that EPA had taken with the other 5 industries that received a further review. We also examined the planning documents for 2 industrial categories—airport deicing and construction and development—that did not go through EPA's 2003-2010 screening and review process but were the subject of regulatory activity during our study period.

To examine limitations to EPA's screening and review process, if any, that could hinder the effectiveness of the effluent guidelines program in advancing the goals of the Clean Water Act, we pursued three separate methodologies: we (1) interviewed a cross section of experts on EPA's effluent guidelines program, (2) surveyed the water quality permit directors of the 46 states that are authorized to issue permits for the National Pollutant Discharge Elimination System (NPDES), and (3) analyzed information about the hazard data sources EPA uses in its screening process.

We identified individuals for possible "expert" interviews by compiling a list of approximately 50 people from a variety of sources relevant to the effluent guideline program, including referrals from EPA, the Association of Clean Water Agencies, and the National Association of Clean Water Agencies and by consulting other knowledgeable individuals, relevant academic literature, and litigation documents. We classified the individuals by their affiliation with a particular stakeholder category (academia, industry, nongovernmental organization, or state and local water quality agencies). We then excluded from consideration 13 individuals for whom we could not obtain contact information. We called or sent an electronic message to those individuals for whom we had contact information to ask if they were familiar with EPA's current effluent guidelines screening and review process. We excluded from consideration those individuals who told us that they were not familiar with these processes, those who could not speak with us during the time frame of our review, and those who said they were not interested in contributing to our review. From our larger list of approximately 50 experts, we selected 22 individuals to interview whom we determined to be experts on the basis of their familiarity with the program and their affiliation with a particular stakeholder category. We conducted 17 interviews including these 22 individuals from February 2012 to April 2012. Six of these interviews were with officials from industry, 4 from academia, 4 from state and local government, and 3 from nongovernmental organizations. In 4 cases, more than one expert participated in an interview. We prepared and asked a standard set of questions about the overall effectiveness of the effluent guidelines program and EPA's use of hazard data, stakeholder input, and information on treatment technology in the screening process. We then reviewed their responses to identify common themes. The sample of experts is a nonprobability sample, and we therefore cannot generalize their opinions across all experts on the effluent guideline program.

To assess the extent to which effluent guidelines might need to be revised, we conducted a web-based survey of state water quality directors, and we statistically analyzed the data. Appendix II presents a complete description of our survey and our data analysis.

To obtain information about an industry that EPA had not analyzed in a further review phase, we selected one of the nine industries that states in our survey said presented a risk to human health or the environment, had treatment technology available to reduce that risk, and warranted revision. We asked officials from the five states whose responses for the metal finishing industry met all three of the above criteria a standard set of questions about the risk the metal finishing industrial category posed, the technology available to mitigate this risk, and the likely effect of a revised effluent guideline.

We further interviewed experts about their views on the adequacy of the hazard data that EPA uses in its screening process—discharge monitoring reports and the Toxics Release Inventory—and whether the experts had suggestions for alternative data sources. We also reviewed EPA's own examinations of the benefits and limitations associated with the two data sources. EPA reports on these examinations of data quality in the technical support documents that accompany its effluent guideline program plans. In addition, we interviewed officials from EPA's Office of Enforcement and Compliance Assurance to learn about the management of the databases that store discharge monitoring data. We also interviewed officials from the Engineering and Analysis Division in EPA's Office of Water about possible effects that incomplete or inaccurate data could have on the screening process. We did not perform an independent assessment of data quality, although we concluded from the information we gathered that the data do have limitations that could affect EPA's screening process.

To examine the actions EPA has taken to address any limitations in its screening and review process, we interviewed effluent guideline program officials from the Engineering and Analysis Division about their plans to modify the biennial screening and review process. We also reviewed papers prepared for the division by a contractor, which describe new sources of data that the division could use to identify industrial categories potentially posing environmental hazards and warranting further review for possible new or revised effluent guidelines. In addition, we interviewed officials from the

Engineering and Analysis Division, the Office of Wastewater Management, and the Office of Enforcement and Compliance Assurance about agency efforts to revise the NPDES permitting process and the database that contains NPDES permit information. We conducted these interviews to determine what steps EPA has taken or could take to use these activities to improve the hazard and treatment technology data available for the screening process.

We conducted this performance audit from September 2011 to September 2012, in accordance with generally accepted government auditing standards. Those standards require that we plan and perform the audit to obtain sufficient, appropriate evidence to provide a reasonable basis for our findings and conclusions based on our audit objectives. We believe that the evidence obtained provides a reasonable basis for our findings and conclusions based on our audit objectives.

APPENDIX II: SURVEY OF STATE WATER QUALITY PERMIT WRITERS AND ANALYSIS OF VIEWS ABOUT WHETHER EPA SHOULD REVISE EFFLUENT GUIDELINES

To assess the extent to which effluent guidelines might need to be revised, and to better understand the reasons for any such revisions, we conducted a web-based survey of state water quality officials, and we statistically analyzed patterns in the survey data. Our analysis identified numerous industries in numerous states for which state officials think that EPA should revise its guidelines. Furthermore, our analysis suggests that a few key factors—particularly, the significance of risk posed by effluent and the availability of pollution control technology—largely influence these officials' views about whether guidelines should be revised. Details about our survey and our data analysis follow.

Questionnaire Design

We designed our survey to ask respondents both (1) whether they thought EPA should revise effluent guidelines for certain industrial categories and (2) whether they thought the major factors that EPA considers when revising effluent guidelines were present for these industrial categories in their state.

We reviewed EPA's 2002 draft *Strategy for National Clean Water Industrial Regulations* and identified the four key factors that the agency uses to determine whether effluent guidelines should be revised. These factors include (1) whether the effluent from a particular industrial category poses a significant risk to human health or the environment; (2) whether technology is available to substantially reduce the risk; (3) whether industry could adopt the technology without experiencing financial difficulty; and (4) whether other factors are present, such as whether current effluent guidelines for that industrial category are difficult to administer and whether revised guidelines could promote innovative regulatory approaches. We summarized these factors, using the exact language from EPA's guidance wherever possible, and wrote survey questions that were simple enough to yield valid responses. We determined that the fourth factor was too complicated to be expressed as a single survey question, and we divided it into two simpler questions. By designing the questionnaire in this way, we sought to increase the reliability of our survey data in two ways: First, asking respondents to assess each of the factors that EPA considers for revision before providing their views about whether EPA should revise effluent guidelines focused their attention on providing an informed opinion. Second, by obtaining data on both the decision-making factors and the need for effluent guideline revisions, we were able to conduct a statistical analysis to identify how these factors appear to influence states' views about the need for guideline revisions.

Our survey was divided into three sections. In the first section, we asked states to respond to a series of questions about each of the five industrial categories that release the greatest amount of toxic effluent in their state. We originally considered surveying states about each of the 58 industrial categories regulated by effluent limitation guidelines. During initial interviews with state officials, however, we determined that this approach would be burdensome and impractical. Therefore, we used data on pollutant discharges from EPA's Toxics Release Inventory and discharge monitoring reports to select the five industries that discharged the greatest amount of toxic effluent in each state in 2010.[1]

For each of these five industrial categories, we asked states six questions, the first five of which pertain to EPA's decision-making factors and the last of which pertains to the need for revised effluent guidelines. The six questions we asked about each industry are as follows:[2]

1. Are the existing effluent guidelines for this industry3 sufficient on their own—that is, without additional water quality-based effluent

limits—to protect your state from significant risks4 to human health or the environment?
2. Is there a technology, process change, or pollution prevention action that is available to this industry that would substantially reduce any risks that remain after the state applies existing effluent limits?
3. Do you think this industry can afford to implement this risk-reducing technology, process change, or pollution prevention action without experiencing financial difficulty?
4. Are the current effluent guidelines for this industry difficult to understand, implement, monitor, or enforce?
5. Do you think the current effluent guidelines for this industry could be revised to promote innovative approaches, such as water quality trading or multimedia benefits?
6. Given your responses to the previous questions, do you think EPA should revise the current effluent guidelines for this industry?

In addition to asking about the top five industrial categories in each state, we asked states about two other sets of industrial categories. First, we asked state officials to list up to three other categories that were not among the top five in their state but for which they thought the effluent guidelines should be revised.

Second, we asked these officials to list up to three categories that are not regulated by effluent guidelines but for which they think EPA should consider developing guidelines.

To be confident that our questions would yield reliable data, we conducted four pretests with state officials. During these pretests, we sought to determine whether the questions were clear, could be reliably answered, and imposed a reasonable burden on respondents.

Survey Respondents

We administered our survey to the directors of the water quality programs in the 46 states that are authorized to implement NPDES. These state officials are largely responsible for issuing permits to industrial facilities and for incorporating effluent guidelines into those permits.

They have regular, firsthand experience with the guidelines, and their experience may supplement EPA's information on effluent. We determined that these officials were therefore sufficiently knowledgeable to answer our survey questions. We obtained a list of these officials and their contact information from EPA and verified this list through Internet searches and phone calls with state officials.

We identified the primary contact for each state but asked these individuals to consult with others in their office to determine the most accurate answer for each survey question.

Survey Administration

We implemented our survey as a web-based questionnaire. We notified the state water quality permit directors in February 2012 of our intent to conduct the survey and requested their participation. We instructed the states on how to access the web-based survey on March 2, 2012. We sent three e-mail reminders and telephoned states that had not responded before we closed the survey in April.

We received responses from 31 of the 46 states, for an overall response rate of 67 percent of states. The survey data are based on responses from 42 individuals in these 31 states.

Because we surveyed state officials only about the industrial categories that discharge the greatest amount of toxic effluent in their state, and because several states did not respond to our survey, the results of our analysis are not generalizable to all industrial categories in all states.

Summary of Survey Responses

To determine the extent to which state officials think that effluent guidelines should be revised, we analyzed the univariate frequencies of responses to our six primary survey questions. We aggregated the survey responses to create industry-by-state cases, such that each case represented the views of a particular state about the guidelines for a particular industrial category in that state.

The completed survey questionnaires from 31 states led to 155 possible state-by-industry cases. Because not all states responded to all of the survey questions, however, we had at most 123 valid cases for analysis, depending upon the survey question. A summary of the responses to these questions appears in table 5.

Table 5. State Officials' Responses to the Key Questions in Our Survey for the Industries Discharging the Greatest Amount of Toxic Effluent in Their State

	Probably yes or definitely yes (percentage)	Probably no or definitely no (percentage)	Don't know or no response (percentage)	Total number of cases[a] (percentage)
Are the existing effluent guidelines for this industry sufficient on their own—that is, without additional water quality-based effluent limits—to protect your state from significant risks to human health or the environment?	51 (42)	69 (57)	2 (2)	122 (100%)
Is there a technology, process change, or pollution prevention action available to this industry that would substantially reduce any risks that remain after the state applies existing effluent limits?	38 (31)	51 (41)	34 (28)	123 (100)
If yes to the previous question: Do you think this industry can afford to implement this risk-reducing technology, process change, or pollution prevention action without experiencing financial difficulty?	31 (82)	2 (5)	5 (13)	38 (100)
Are the current effluent guidelines for this industry difficult to understand, implement, monitor, or enforce?	30 (24)	91 (74)	2 (2)	123 (100)
Do you think the current effluent guidelines for this industry could be revised to promote innovative approaches, such as water quality trading or multimedia	44 (36)	59 (48)	19 (16)	122 (100)

	Probably yes or definitely yes (percentage)	Probably no or definitely no (percentage)	Don't know or no response (percentage)	Total number of cases[a] (percentage)
benefits?				
Given your responses to the previous questions, do you think EPA should revise the current effluent guidelines for this industry?	63 (51)	60 (49)	0 (0)	123 (100%)

Source: GAO analysis of survey data.

[a] This column represents all cases for which the survey respondent selected one of the response options, which included "don't know/no response." It does not include responses from individuals who skipped the question entirely.

These tabulations indicate that a substantial number of cases exist for which states thought that EPA should revise effluent guidelines and also for which they perceived that one or more of EPA's decision-making factors were present. In 51 percent (63 of 123 cases), state officials said that EPA should revise the effluent guidelines for the corresponding industry. With regard to whether the key decision-making factors were present, state officials reported that effluent posed a significant risk in 57 percent of cases, that technology was available in 31 percent of cases, that the guidelines were difficult to administer in 24 percent of cases, and that revised guidelines could promote innovative approaches in 36 percent of cases.

We had far fewer responses to our question about whether industry could adopt technology without experiencing financial difficulty because that question was applicable only if the respondent said such technology was available. Among these cases, state officials reported that the technology would not cause financial hardship to the industry in 82 percent of cases (31 of 38 cases).

We repeated this analysis after removing the 29 cases representing the three industrial categories whose effluent guidelines are in revision, leaving at most 96 cases for analysis, depending upon the question. Of the remaining cases, state officials said that EPA should revise the effluent guidelines for a substantial percentage of them; they also said that key decision-making factors were present in a substantial percentage of cases. For example, in 46 percent of these cases, state officials said that EPA should revise the effluent guidelines for the corresponding industry.

Relationships between Key Decision-Making Factors and States' Views about Whether EPA Should Revise Effluent Guidelines

We compared state officials' views about whether effluent guidelines should be revised with their views of each of the factors that EPA uses when considering guideline revisions. For three of the four factors, our results show that when state officials perceived the factor to be present, they were significantly more likely to think that EPA should revise the effluent guidelines for the corresponding industrial category. (We had too few cases with valid responses to the survey question about cost to determine whether that factor was significantly associated with views about guideline revisions.) The risk posed by effluent and the availability of technology were the strongest predictors of states' views about the need for guideline revisions. In particular, we found the following:

- When state officials perceived effluent from a particular industrial category to pose a significant risk, they were 3.8 times more likely to think that EPA should revise the guidelines for that category than when they did not perceive the effluent to pose a significant risk. Specifically, among the cases in which state officials perceived effluent to pose a significant risk, they thought the effluent guidelines should be revised 75 percent of the time (52 of 69 cases), compared with 20 percent of the time (10 of 51 cases) when they thought the effluent did not pose a significant risk.
- When state officials perceived technology to be available to substantially reduce the risk for a particular industrial category, they were 4.3 times more likely to think that EPA should revise the guidelines for that category than when they did not perceive technology to be available. Specifically, among the cases in which these officials perceived technology to be available, they thought EPA should revise the effluent guidelines 84 percent of the time (32 of 38 cases), compared with 20 percent (10 of 51 cases) when they thought that technology was not available.
- When state officials thought that other factors were present for a particular industrial category, they were 2.3 times more likely to think that EPA should revise the guidelines than when they did not think these factors were present. "Other factors" refers to either that the current guidelines were difficult to understand, implement, monitor, or enforce or that revised guidelines could promote innovative

approaches. Specifically, when state officials thought that such other factors were present, they thought that EPA should revise its effluent guidelines 70 percent of the time (43 of 61 cases), compared with 30 percent of the time (18 of 60 cases) when they thought these factors were not present.

Table 6 presents the complete results of these bivariate comparisons. We excluded one of the factors from the discussion above—namely, whether the industry could afford to implement the technology, process change, or pollution prevention action—because the responses to this question applied only to the subset of cases for which such a technology, change, or action was available, only 33 of which provided a yes or no response. In 87 percent of those cases in which the technology was perceived to be affordable (27 of 31 cases), state officials said that EPA should revise its guidelines for the corresponding industry. We repeated this analysis after removing the 29 cases representing the two industrial categories whose effluent guidelines EPA is already revising. We found that, even after removing these cases, the same three factors retained a significant relationship with state officials' views about whether effluent guidelines should be revised. This result indicates that these key decision-making factors appear to influence state officials' views even for industrial categories whose guidelines EPA is not already revising.

Table 6. State Officials' Views about Whether EPA Should Revise the Effluent Guidelines for the Industries Discharging the Greatest Amount of Toxic Effluent in Their State, by the Four Factors EPA Considers When Deciding Whether to Revise Effluent Guidelines

Do you think EPA should revise the current effluent guidelines for this industry?			
	Probably yes or definitely yes (percentage)	Probably no or definitely no (percentage)	Total number of cases[a] (percentage)
Are the existing effluent guidelines for this industry sufficient on their own—that is, without additional water-quality-based effluent limits—to protect your state from significant risks to human health or the environment?			
Probably yes or definitely yes	10 (20)	41 (80)	51 (100%)
Probably no or definitely no	52 (75)	17 (25)	69 (100)
Total	62 (52)	58 (48)	120 (100)

Table 6. (Continued)

Do you think EPA should revise the current effluent guidelines for this industry?			
	Probably yes or definitely yes (percentage)	Probably no or definitely no (percentage)	Total number of cases[a] (percentage)
Is there a technology, process change, or pollution prevention action available to this industry that would substantially reduce any risks that remain after the state applies existing effluent limits?			
Probably yes or definitely yes	32 (84)	6 (16)	38 (100)
Probably no or definitely no	10 (20)	41 (80)	51 (100)
Total[b]	42 (47)	47 (53)	89 (100)
If yes to the previous question: Do you think this industry can afford to implement this risk-reducing technology, process change, orpollution prevention action without experiencing financial difficulty?			
Probably yes or definitely yes	27 (87)	4 (13)	31 (100)
Probably no or definitely no	1 (50)	1 (50)	2 (100)
Total[c]	28 (85)	5 (15)	33 (100)
Are the current effluent guidelines for this industry difficult to understand, implement, monitor, or enforce or could the current effluentguidelines for this industry be revised to promote innovative approaches?[d]			
Probably yes or definitely yes	43 (70)	18 (30)	61 (100)
Probably no or definitely no	18 (30)	42 (70)	60 (100)
Totala	61 (50)	60 (50)	121 (100%)

Source: GAO analysis of survey data.

[a] This column represents all cases for which the survey respondent selected one of the response options, which included "don't know/no response," for both the question on whether EPA should revise its effluent guidelines and the question on whether a given factor was present. It does not include responses from individuals who skipped the questions entirely.

[b] Cases were excluded from this analysis if the response to either question in the cross-tabulation was "Don't know or no response."

[c] This question pertained only to the cases for which respondents answered that a technology, process change, or pollution prevention action was available.

[d] This category combines two survey questions. Cases in this category were coded as "probably yes or definitely yes" if that response was given to either of the two questions.

Decision Tree of States' Views about Whether EPA Should Revise Effluent Guidelines

To understand how the various decision-making factors interact to influence states' views about the need for revised effluent guidelines, we used the data from our survey to conduct decision-tree analysis. We developed the decision tree by splitting the data into smaller and smaller subgroups according to whether state officials perceived each of the factors to be present for a particular industrial category. Beginning with the first factor, risk, we divided the cases into subgroups, depending upon whether state officials perceived the effluent from the particular industry to pose a significant risk to human health or the environment. For each of these subgroups, we tabulated the number of cases in which state officials said the effluent guidelines should be revised, compared with the number of cases in which they said the guidelines should not be revised. We then split these subgroups again, according to whether state officials thought that technology was available to substantially reduce the risk. This split resulted in further subgroups. We continued splitting the data into smaller and smaller subgroups by next assessing state official's views of the cost of technology and finally assessing their views on the presence of other factors. At each step, we stopped splitting the data if (1) the original group had fewer than 10 cases, (2) the resulting subgroups did not differ significantly in terms of the percentages of respondents who said that EPA should revise the guidelines; or (3) the resulting subgroups tended to support the same conclusion as to whether EPA should revise the guidelines. We examined the cases terminating in each of the branches and found that the overall decision tree was based on a broad variety of industries and states. The resulting decision tree, which is shown in figure 5, has four splits and six branches.

The decision tree illustrates how the key decision-making factors collectively predict states' views about whether EPA should revise effluent guidelines, and it corroborates the reliability of our survey data. Overall, when the risk of effluent was perceived to be significant and technology was perceived to be available, state officials overwhelmingly thought the corresponding effluent guidelines should be revised.

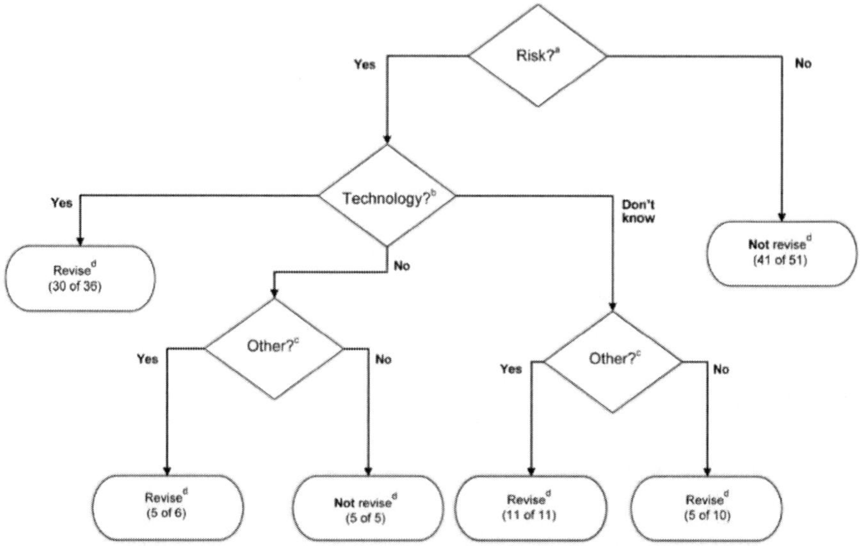

Source: GAO analysis of data from survey of state water quality officials.

Note: This analysis is based on 119 industry-by-state cases from our survey of state water quality permit writers. Each case represents the views of a single state about a single industry in that state.

[a] Whether the state official views the effluent from a particular industry to pose a significant risk to human health or the environment, according to their response to the first question on our survey.

[b] Whether the state official views technology to be available to substantially reduce the risk to human health or the environment, according to their response to the second question on our survey.

[c] Whether the state official views other factors to be present—such as current guidelines difficult to enforce or revised guidelines able to promote innovative approaches—according to the fourth and fifth questions in our survey.

[d] Based on responses to the sixth question on our survey, as to whether state officials think EPA should revise the effluent guidelines for a particular industrial category. The tabulations in parentheses represent the number cases in which state officials answered yes and no, respectively, to this question for each branch of the decision tree.

Figure 5. Decision Tree of State Officials' Views of Whether EPA Should Revise Effluent Guidelines for Specific Industrial Categories.

Even when technology was not perceived to be available, many states still thought the guidelines should be revised if they thought that other factors were present. In particular, in three scenarios, corresponding to three branches of the decision tree, state officials generally said that effluent guidelines should be revised:

- When state officials thought that effluent from an industrial category poses a significant risk to human health or the environment and when they thought technology was available to substantially reduce that risk, they generally said that EPA should revise the effluent guidelines. In such instances, they thought that EPA should revise the effluent guidelines 83 percent of the time (in 30 of 36 cases). This scenario is illustrated by the far left branch of the decision tree.
- When state officials thought that effluent from an industrial category poses a significant risk, they generally thought that EPA should revise the effluent guidelines even when they perceived that technology was not available—as long as they perceived other factors to be present. In such instances, they thought that EPA should revise its effluent guidelines 83 percent of the time (5 of 6 cases). This scenario is illustrated by the second-to-left branch of the decision tree.
- When state officials thought that effluent from an industrial category poses a significant risk, they generally thought that EPA should revise the effluent guidelines even when they did not know if technology was available—as long as they perceived other factors to be present. In such instances, these officials thought EPA should revise its effluent guidelines 100 percent of the time (11 of 11 cases). This scenario is illustrated by the branch of the decision tree in the third column from the right.

By contrast, in two scenarios, state officials thought EPA should not revise the guidelines. In the primary scenario, officials did not perceive the effluent to pose a significant risk, although officials also thought that guidelines should not be revised when the risk was significant but neither technology nor other factors were present. In particular, our decision tree identified the following two scenarios:[5]

- When state officials did not think the effluent from a particular industrial category posed a significant risk to human health or the environment, they generally thought that EPA should not revise the corresponding effluent guidelines. In these instances, state officials thought that EPA should not revise the guidelines 80 percent of the time (41 of 51 cases). This scenario is illustrated by the branch of the decision tree on the far right.

- When state officials thought the effluent from a particular industrial category posed a significant risk but that technology was not available and other factors were not present, they generally said that EPA should not revise the effluent guidelines for that industry. In such instances, state officials thought that EPA should not revise the guidelines 100 percent of the time (5 of 5 cases). This scenario is illustrated by the branch of the decision tree in the third column from the left.

Industrial Categories for Which States Thought Effluent Guidelines Should Be Revised

Corresponding to this decision tree, we further examined the data to identify specific industrial categories that presented the strongest evidence for needing to be revised.

Because the significance of risk and the presence of technology are the two primary decision-making factors, we selected the 30 cases for which states said these two factors were present and for which they said effluent guidelines should be revised. These cases fall into the far left branch of the decision tree in figure 5.

These 30 cases represent 14 industrial categories: canned and preserved seafood processing; cement manufacturing; coal mining; fertilizer manufacturing; meat and poultry products; metal finishing; metal molding and casting; oil and gas extraction; ore mining and dressing; petroleum refining; pulp, paper, and paperboard; steam electric power generation; sugar processing; and timber products processing. We added industries that state officials cited in the second section of our survey, in which we asked them to identify industries that were not among the top five dischargers in their state. This addition lengthened the list by 22 cases, representing 7 additional industrial categories: centralized waste treatment, dairy products processing, electrical and electronic components, electroplating, grain mills manufacturing, landfills, and pharmaceutical manufacturing. In total, therefore, we identified 52 cases representing 21 industrial categories for which state officials thought effluent guidelines should be revised. Of these 52 cases, 39 represent industrial categories whose guidelines EPA is not already revising.

APPENDIX III: ADDITIONAL DETAILS ON INDUSTRIAL CATEGORIES WITH EFFLUENT GUIDELINES

EPA has promulgated effluent guidelines for 58 industrial categories beginning in the mid-1970s. EPA has also revised the guidelines for most of those industries, although many have not been revised in recent years. As described elsewhere in this report, EPA uses a screening process to determine which categories may warrant further review and possible revision. According to our analysis, since EPA began using its current screening process in 2003, more than half the industrial categories with effluent guidelines did not advance beyond the screening phase in any year from 2003 to 2010 because, during a given 2-year screening cycle, the relative toxicity of their pollutant discharges did not put them among the top 95 percent of discharge hazard. Table 7 provides further information on the industrial categories, including the year their effluent guidelines were first promulgated, the year the guidelines were most recently revised, and the year(s) in 2004 through 2010 when their hazard ranking scores came within the top 95 percent.

Table 7. Years Effluent Guidelines Were Promulgated and Revised for Industrial Categories and Years the Categories Were in Hazard, 2004-2010 the Top 95 Percent of Total Repo

			Year(s) the industrial category was in the top 95 percent of total hazard			
Industrial category	Year promulgated	Year most recently revised	2010	2008	2006	2004
Airport deicing	2012	Not revised				
Aluminum forming	1983	1988				
Asbestos manufacturing	1974	1995				
Battery manufacturing	1984	1986				
Canned and preserved fruits and vegetables processing	1974	1995				
Canned and preserved seafood processing	1974	1995				
Carbon black manufacturing	1978	1995				
Cement manufacturing	1974	1995	X			
Centralized waste treatment	2000	2003		X		
Coal mining	1985	2002	X			

Table 7. (Continued)

			Year(s) the industrial category was in the top 95 percent of total hazard			
Industrial category	Year promulgated	Year most recently revised	2010	2008	2006	2004
Coil coating	1982	2007				
Concentrated animal feeding operations	2003	2008				
Concentrated aquatic animal production	2004	Not revised				
Construction and development	2009	Not revised				
Copper forming	1983	1986				
Dairy products processing	1974	1995				
Electrical and electronic components	1983	1985				
Electroplating	1981	1986				
Explosives manufacturing	1976	1995				
Ferroalloy manufacturing	1974	1995				
Fertilizer manufacturing	1974	1995	X	X	X	X
Glass manufacturing	1974	1995				
Grain mills	1974	1995				
Gum and wood chemicals manufacturing	1976	1995				
Hospital	1976	1995				
Ink formulating	1975	1995				
Inorganic chemicals manufacturing	1982	1984	X	X	X	X
Iron and steel manufacturing	1982	2005				
Landfills	2000	2000	X			
Leather tanning and finishing	1982	1996				
Meat and poultry products	2004	Not revised				
Metal finishing	1983	1986				
Metal molding and casting	1985	1986	X			
Metal products and machinery	2003	Not revised				
Mineral mining and processing	1975	1995	X			
Nonferrous metals forming and metal powders	1985	1989				

			Year(s) the industrial category was in the top 95 percent of total hazard			
Industrial category	Year promulgated	Year most recently revised	2010	2008	2006	2004
Nonferrous metals manufacturing	1984	1990	X	X	X	X
Oil and gas extraction	1979	2012a	X			
Ore mining and dressing	1982	1988	X	X	X	X
Organic chemicals, plastics, and synthetic fibers	1987	1993	X	X	X	X
Paint formulating	1975	1995				
Paving and roofing materials (tars and asphalt)	1975	1995				
Pesticide chemicals	1978	1998	X	X	X	
Petroleum refining	1982	1985	X	X	X	X
Pharmaceutical manufacturing	1983	2003				
Phosphate manufacturing	1974	1986				X
Photographic	1976	Not revised				
Plastics molding and forming	1984	1985	X		X	
Porcelain enameling	1982	1985				
Pulp, paper, and paperboard	1998	2012[a]	X	X	X	X

Source: GAO analysis of EPA documentation.

Notes: In its screening phase, EPA ranks some industrial categories that are not subject to existing effluent guidelines and are therefore not included in this table. When EPA revised the effluent guideline for an industrial category, it may have revised just a portion of the guideline. For example, EPA may have added pollutants or changed the limits for a particular industrial category or added a new subcategory. In some cases, EPA may have made revisions that did not affect the stringency of the effluent guidelines. With the exception of three 2012 revisions, we did not determine the nature of the revisions shown in this table.

[a] The revisions to these industrial categories did not increase the stringency of the effluent guidelines.

End Notes

[1] The Federal Water Pollution Control Act Amendments of 1972, Pub. L. No. 92-500, § 2, 86 Stat. 816, codified as amended at 33 U.S.C. §§ 1251-1387 (2012) (commonly referred to as the Clean Water Act). For consistency throughout this report, we refer to the statute and its amendments as the Clean Water Act.

[2] In addition to the 46 states, the territory of the U.S. Virgin Islands is authorized to issue NPDES permits. The remaining territories and the District of Columbia are not authorized to issue these permits.

[3] A Senate committee report explaining the addition of this provision to the Water Quality Act of 1987, noted in part, "[g]uidelines are required for any category of sources discharging significant amounts of toxic pollutants. In this use, 'significant amounts' does not require the Administrator to make any determination of environmental harm; any nontrivial discharges from sources in a category must lead to effluent guidelines." S. Rep. 99- 50 at 24-25 (1985). See also 69 Fed. Reg. 53,707 (Sept. 2, 2004).

[4] Consent decrees are settlement agreements signed by the parties and entered, or approved, by a court; they are therefore enforceable by the courts.

[5] EPA stayed a portion of the guideline for the construction and development industrial category that established a numerical effluent limitation for turbidity, but other portions of the guidelines remain in effect. See 75 Fed. Reg. 68215 (Nov. 5, 2010), 40 C.F.R. pt. 450.

[6] As of August 2012, EPA had not published a preliminary effluent guidelines program plan for 2011.

[7] Throughout this report, we use the term effluent guidelines to refer to effluent guidelines and pretreatment standards collectively. Some industrial categories are made up of direct dischargers, some of indirect dischargers, and some have a mix of both. Thirty-four of the 58 industrial categories are regulated by both effluent guidelines and pretreatment standards because these categories comprise both direct and indirect dischargers. Twenty-three industrial categories are regulated by effluent guidelines but not pretreatment standards, while the electroplating industrial category is regulated by pretreatment standards but not by effluent guidelines.

[8] Wastewater treatment plants generally must have a NPDES permit to operate.

[9] EPA regulations require permit writers to document the reasoning behind a facility's permit. A fact sheet is required to accompany the permit for facilities designated by EPA regional administrators and the permitting authorities to be major dischargers. A statement of basis is required for permits issued to all other facilities, which EPA considers minor facilities. For purposes of this report, we refer to both fact sheets and statements of basis as "fact sheets." See EPA, NPDES Permit Writers Manual at 11-10, 40 C.F.R. §§ 124.7, 124.8, 124.56, 123.25 (2012).

[10] EPA and the states are making a transition from one national database, known as the Permit Compliance System, to another known as the Integrated Compliance Information System: NPDES. The states are divided in their use of the two databases. Consequently, two databases contain discharge-monitoring reports. In our report, however, we refer to them collectively as "the database."

[11] Specifically, certain facilities that manufacture, process, or otherwise use any of the listed individual chemicals and chemical categories are required to report annually to EPA and their respective state those chemicals used above threshold quantities, the amounts released to the environment, and whether the releases entered the air, water, or soil. 42 U.S.C. § 11023 (2012).

[12] Biological oxygen demand is a measure of the oxygen used during decomposition of organic material over a specified period (usually 5 days) in a wastewater sample; it represents the readily decomposable organic content of wastewater.

[13] A measure of filterable solids present in a sample, as determined by the method specified in 40 C.F.R. pt. 136.

[14] Fecal coliform are bacteria whose presence indicates that water may be contaminated by human or animal wastes.

[15] A measure of the hydrogen ion concentration of water or wastewater expressed as the negative logarithm of the hydrogen ion concentration in milligrams per liter. A pH of 7 is neutral, a pH less than 7 is acidic, and a pH greater than 7 is basic.

[16] See Clean Water Act §307(a)(1), codified at 33 U.S.C. §1317(a)(1) (2011); see also 33 U.S.C. § 1362(13) (defining toxic pollutant). The list appears in the Code of Federal Regulations at 40 C.F.R. § 401.15. The committee report list was developed from a 1976 consent decree signed with the Natural Resources Defense Council, an environmental group, to resolve litigation that, among other things, sought to compel EPA to expand the list of toxic pollutants promulgated under the act. The consent decree was entered in NRDC et al. v. Train, 6 ELR 20588, (D.D.C. 1976). The statute authorizes EPA to revise the list.

[17] Human-made halomethanes are most notably used as refrigerants, solvents, propellants, and fumigants.

[18] Chemical oxygen demand is a measure of the oxygen-consuming capacity of inorganic and organic matter present in wastewater.

[19] See, e.g., Senate Consideration of the Report of the Conference Committee, October 4, 1972 (Statement of Sen. Muskie), reprinted in Cong. Research Serv., A Legislative History of the Water Pollution Control Act Amendments of 1972, at 169-70 (1978); S. Comm. on Public Works, Water Pollution Control Act Amendments of 1972, S. Rep. No. 92-414, at 50-51 (1971), reprinted in Cong. Research Serv., A Legislative History of the Water Pollution Control Act Amendments of 1972, at 7669-70 (1978).

[20] EPA is required to issue both effluent guidelines and effluent limitations. The agency issues regulations that simultaneously address both of these and therefore it does not distinguish between the reviews required for the guidelines and for the limitations.

[21] See also 68 Fed. Reg. 75,515, 75,519 (Dec. 31, 2003) in which EPA stated, "EPA articulated an early form of this evolving analytical framework in the draft Strategy for National Clean Water Industrial Regulations ('draft Strategy'), which EPA hope[d] to finalize concurrently with the Effluent Guidelines Program Plan in 2004."

[22] Our Children's Earth (OCE) Found. v. EPA, 527 F.3d 842, 851 (9th Cir. 2008), rehearing 506 F.3d 781 (9th Cir. 2007), on appeal from 2005 U.S. Dist. Lexis 45716 (N.D. Cal. 2005).

[23] EPA's draft 2002 strategy stated that it would consider the risks to human health or the environment. According to a senior effluent guidelines program official, however, the agency's screening process includes a relative hazard assessment rather than a risk assessment. According to EPA, once an industrial category has been identified as posing a significant hazard on the basis of the screening analysis—and before initiating an effluent guideline rule making—the agency may then conduct a study of the industrial category to determine the risks imposed on human health and the environment.

[24] The steam electric power-generating industry produces electric power by means of steam generated from fossil fuels, such as coal, oil, and natural gas, or nuclear fuels.

[25] As explained above, an industrial direct discharger is required to have an NPDES permit regardless of whether there are effluent guidelines for the industry. NPDES permits require monitoring for specific pollutants to determine compliance with permit limits. Some industries may also be subject to requirements under another EPA program to report toxic releases to the Toxics Release Inventory. These requirements are independent of whether an industry is regulated by effluent guidelines.

[26] According to EPA, the doubling in the number of industrial categories in 2010 resulted from the removal of the steam electric power-generating category from the ranking process after

EPA decided to revise its effluent guidelines. Previously, that industrial category alone constituted up to 73 percent of the total toxic weighted pound equivalent. When EPA removed that category from its hazard ranking, many other industries with smaller hazard ranking scores moved into the top 95 percent.

[27] The industrial coding schemes that EPA uses are the Department of Labor's Standard Industrial Classification codes, created in the 1930s and the North American Industry Classification System, which was created through a cooperative effort of Canadian and Mexican government agencies and the U.S. Office of Management and Budget.

[28] Under the Paperwork Reduction Act, EPA can contact—with a survey or questionnaire—up to 9 entities without first obtaining approval from the Office of Management and Budget. If EPA decides to contact 10 or more entities, the agency must prepare an Information Collection Request. This request describes the information to be collected, gives the reasons the information is needed, and estimates the time and cost for respondents to answer the request. The office reviews the request and determines if the request is approved or disapproved, or it defines conditions to be met for approval.

[29] EPA also announced in October 2011 the initiation of a new effluent guideline rulemaking process for shale gas extraction. The agency decided to undertake the rulemaking on the basis of stakeholder concerns about the industrial category without going through a further review phase. The agency plans to propose new standards in 2014. In addition, EPA conducted the further review phase of the airport deicing industrial category prior to our 2003 time frames. The agency issued effluent guidelines for the category in May 2012.

[30] Generally, permits are to establish limitations for those pollutants reasonably expected to be present in wastewater with potential to cause or contribute to an excursion above a water quality standard. For an industry with an effluent guideline, the guideline specifies which pollutants must, at minimum, be included in the permit. For other industries, the permit writer uses information provided on the permit application, as well as other sources, to determine which pollutants may be present in wastewater and warrant a limitation. In addition, permits may include water-quality-based limits derived from the standards for the water body into which the effluent is discharged.

[31] In addition, EPA has identified some limitations in the discharge monitoring report data that may cause the agency to overestimate the hazard presented by an industrial category. For example, many facilities do not report average quantities for specific pollutants, in which case, EPA has to base its estimates on the maximum or other amount discharged, which could lead to overestimating a facility's actual discharges.

[32] And, as we described above, after identifying the industrial categories contributing to the top 95 percent of hazard, EPA may use other factors to exclude additional industrial categories before beginning the further review phase. Therefore, even those categories within the top 95 percent do not necessarily receive the further review examining the availability of treatment technology.

[33] 68 Fed. Reg. 75515, 75521 (Dec. 31, 2003) ("EPA found that it was much more difficult than anticipated to gather the data needed to perform a meaningful screening-level analysis of the availability of treatment or process technologies that might reduce hazard or risk beyond the performance of technologies in place.").

[34] Environmental Protection Agency, Factor 2 Analysis: Technology Advances and Process Changes (Washington, D. C.: December 2003).

[35] Notably, of the six experts we interviewed from industry, only one suggested additional sources.

[36] Air pollution regulations can be relevant in that they may cause a shift in pollutants from air emissions to wastewater or sludge. For example, EPA's ongoing effort to revise the effluent guidelines for the steam electric power-generating industry is in part a response to changes in the industry's wastewater as the plants installed scrubber equipment that uses water to remove pollutants to comply with air pollution regulations to control sulfur dioxide.

[37] As of August 2012, the Office of Management and Budget is reviewing EPA's draft electronic reporting rule. Accordingly, EPA has not yet proposed the rule.

[38] Five experts said that EPA should not consider technology earlier in its screening phase, and two did not provide their opinions.

End Notes for Appendix II

[1] Although these data have certain limitations, which we describe elsewhere in this report, we determined they were sufficiently reliable for the purpose of selecting industries on which to focus our survey questions.

[2] Because state officials might not have complete information about all the key factors for each of the five industries, we allowed them to report their level of certainty in their response by answering each of the questions with either definitely yes, probably yes, probably no, definitely no, or don't know or no response. We collapsed these five categories into three categories—yes, no, and don't know or no response—for subsequent analysis.

[3] In the online version of the questionnaire, we customized the survey questions by inserting the name of each of the specific industries for each state.

[4] We based our survey questions on EPA's draft 2002 strategy, which uses the term risk rather than the term hazard. To be consistent with the precise wording of the survey questions, we use the term risk when describing the survey results. Elsewhere in this report, we use the term hazard in accordance with our discussions with EPA officials about the agency's screening process, in which contaminants are first assessed for hazard and then assessed for risk.

[5] Responses were evenly split when risk was perceived to be present, state officials were uncertain whether technology was available, and they did not report that other factors were present. This split is illustrated by the branch of the decision tree in the second column from the right.

In: Water Pollution Controls
Editor: Julia Crawford
ISBN: 978-1-62417-441-4

Chapter 2

CLEAN WATER ACT AND POLLUTANT TOTAL MAXIMUM DAILY LOADS (TMDLS)*

Claudia Copeland

SUMMARY

Section 303(d) of the Clean Water Act (CWA) requires states to identify waters that are impaired by pollution, even after application of pollution controls. For those waters, states must establish a total maximum daily load (TMDL) of pollutants to ensure that water quality standards can be attained. A TMDL is both a quantitative assessment of pollution sources and pollutant reductions needed to restore and protect U.S. waters and a planning process for attaining water quality standards. Implementation of section 303(d) was dormant until states and the Environmental Protection Agency (EPA) were prodded by lawsuits. The program has been controversial, in part because of requirements and costs faced by states to implement this 40-year-old provision of the law, as well as industries, cities, farmers, and others who may be required to use new pollution controls to meet TMDL requirements.

Despite controversies, the TMDL program has become a core element of overall efforts to protect and restore water quality. States and EPA develop several thousand TMDLs annually, but many more need to be completed. The most recent information indicates that over 41,000

* This is an edited, reformatted and augmented version of a Congressional Research Service publication, CRS Report for Congress R42752, prepared for Members and Committees of Congress, from www.crs.gov, dated September 21, 2012.

waterbodies do not meet water quality standards and need a TMDL to initiate corrective measures.

The 303(d) program has evolved, and especially during the last decade, EPA and states have addressed more complex issues, including TMDLs involving both point (direct discharges) and nonpoint sources (diffuse discharges) such as stormwater; TMDLs for less-traditional causes of impairment such as ocean acidification and climate change; TMDLs for pollutants such as mercury that involve coordination among water, air, and other environmental programs; and multi-jurisdictional TMDLs.

The largest multi-jurisdictional TMDL, for the Chesapeake Bay watershed, has drawn considerable attention. It was developed by EPA and was necessitated because previous largely voluntary restoration efforts by the Bay jurisdictions were insufficient to attain water quality standards.

It addresses all segments of the Bay and its tidal tributaries that are impaired from discharges of nitrogen, phosphorus, and sediment, with a goal of having TMDL implementation measures in place by 2025. The Chesapeake Bay TMDL has a number of novel elements, including Watershed Implementation Plans in which the jurisdictions identify specific measures to achieve needed pollutant reductions, and biennial reports to the public on progress in implementation. The Bay TMDL has been controversial with a number of groups concerned about the costs of implementation and the likely mandatory nature of many of EPA's and states' actions. EPA's authority to develop the TMDL has been challenged in a lawsuit.

When a TMDL is developed, implementation is a major uncertainty. First, section 303(d) does not require implementation, and states' strategies for implementation vary widely. Only a few have laws requiring implementation plans, while many others rely on less structured policies. Second, a number of barriers to implementation can be identified.

The most prominent is insufficient funding, but technical impediments such as insufficient scientific data also are a challenge. At the same time, factors that may aid effective implementation can be identified, including active involvement of stakeholders and governments, and adequate resources.

The TMDL program is in a period of transition and increasingly is addressing new challenges— more complex TMDLs, larger scale impairments, and nonpoint sources. Other than recent oversight hearings on the Chesapeake Bay TMDL, Congress has not shown active interest in the TMDL program for more than a decade. Some stakeholders, especially states, believe that several issues present Congress with an opportunity to examine the TMDL provisions of the CWA.

Background

The Clean Water Act (CWA) contains a number of complex and interrelated elements of overall water quality management. Foremost is the requirement in section 303 that states establish ambient water quality standards for water bodies, consisting of the designated use or uses of a water body (e.g., recreational, public water supply, or industrial water supply) and the water quality criteria which are necessary to protect the use or uses. Standards are then used to determine which waters must be cleaned up, how much effluent may be discharged, and what is needed for protection. Through permitting, states or the Environmental Protection Agency (EPA) impose wastewater discharge limits on individual industrial and municipal facilities to ensure that water quality standards are attained. However, Congress recognized in the act that, in many cases, pollution controls implemented by industry and cities would be insufficient to attain and maintain water quality standards, due to pollutant contributions from other unregulated sources.

Under CWA section 303(d), states must identify waters for which discharge limits specified in permits are not stringent enough to achieve established water quality standards, after implementation of technology-based controls by industrial and municipal dischargers. For each waterbody or segment, states[1] are required to set a total maximum daily load (TMDL) of pollutants at a level that ensures that applicable water quality standards can be attained and maintained. A TMDL is essentially a pollution budget, a quantitative estimate of what it takes to achieve state water quality goals, setting the maximum amount of pollution a waterbody can receive without violating water quality standards, including a margin of safety to account for seasonal variations and uncertainty between pollutant loads and the quality of receiving waters. As such, TMDLs provide a scientific calculation of how much pollutant loads need to be reduced to meet those standards.

A TMDL is both a planning process for attaining water quality standards and a quantitative assessment of problems, pollution sources, and pollutant reductions needed to restore and protect a river, stream, or lake. TMDLs may address all pollution sources and allocate needed pollutant reductions among categories of sources that contribute to the water quality impairment, including point sources, such as municipal sewage treatment or industrial plant discharges (wasteload allocation or WLA); and nonpoint sources (load allocation or LA), such as runoff from roads, farm fields, and forests, atmospheric deposition, naturally occurring sources, and background sources of the pollutant.

The goal of the TMDL is to eliminate an impairment, not meet a pollutant limit for its own sake. The TMDL itself does not establish new regulatory controls on sources of pollution, and it does not set discharge limits. Nor is it self-implementing. However, when TMDLs are established, municipal and industrial wastewater treatment plants may be required to install new or improved pollution control technology. For waters impaired by point source discharges, TMDLs are enforced through revisions to existing CWA permits, which include the pollutant limits and a schedule for compliance. For waters impaired by nonpoint source runoff, because there are no federal controls over these sources (there is no CWA permit requirement as there is for point sources), the primary implementation tools are state-run nonpoint source management programs, coupled with state, local, and federal land management programs and authorities, and financial assistance and incentive programs. States can but are not required to regulate nonpoint sources to achieve goals set out in a TMDL. For example, farmers or ranchers may be asked by states to use alternative methods in their operations to prevent fertilizers and pesticides from reaching streams, and they may receive funds to help them install on-farm pollution management systems or practices.

EPA has specific responsibilities under section 303(d). First, EPA is required to review and approve a state's list of impaired waters and TMDLs. Second, if a state fails to identify impaired waters and develop a TMDL where one is needed, the CWA requires EPA to develop an impaired waters list for the state and make its own TMDL determination. EPA is not authorized to implement a TMDL.

IMPLEMENTATION OF SECTION 303(D)

EPA acknowledges that a vigorous TMDL program is needed because significant water quality problems persist in the nation's waters, nearly 40 years after enactment of the CWA. An estimated 40% of the nation's waters assessed by states do not meet water quality standards.[2] TMDLs are one element of state water quality management programs. Other activities include standard setting, monitoring, permitting, and enforcement, and all must be integrated with the TMDL program. Most states have lacked the resources to do TMDL analyses, which involve complex assessment in order to ascribe and quantify environmental effects from particular discharge sources. Baseline water quality monitoring data for the analyses (to identify impaired waters and pollution sources) have been limited. EPA has both been reluctant to intervene in the states and also lacked resources to do so itself. Thus, there was little initial implementation of the provision enacted in 1972. Only in 1992 did EPA issue regulations requiring states, every two years, to list waters that do not attain water quality standards and establish TMDLs to restore water quality.

Responding to the failure of states and EPA to meet these requirements, however, environmental groups filed lawsuits in more than three dozen states to compel compliance with the law's requirements. Environmentalists see implementation of section 303(d) as important both to achieving the overall goals and objectives of the act and to pressuring EPA and states to address nonpoint and other sources that are responsible for many water quality impairments nationwide but have not been well controlled. Of the suits tried or settled, nearly two dozen resulted in court orders requiring expeditious TMDL development by states or EPA.

The TMDL litigation falls into three general categories: (1) challenges intended to compel EPA to step in to fulfill TMDL requirements where a state has failed partially or completely to do so; (2) challenges to EPA's listing of impaired waters, TMDL approval decisions, or EPA's promulgation of TMDLs; and (3) challenges to the substance or content of TMDLs.[3]

Because of the lawsuits and existing requirements of the law, in 1997, EPA issued interpretive guidance which for the first time called on states to develop long-term schedules for implementing TMDLs. There is neither a CWA nor a regulatory deadline for states to develop TMDLs. Under the 1997 guidance, EPA recommends that states establish TMDLs in order to meet water quality standards within 8 to 13 years of listing the water as impaired.[4] Development of TMDLs has increased (since 1996, states and EPA have established more than 48,000), but many more remain to be completed. The most recent state 303(d) lists, most of them submitted in 2008 and 2010, identified over 41,000 waterbodies as not meeting water quality standards and in need of a TMDL, affecting more than 300,000 miles of rivers and shorelines and 5 million acres of lakes and resulting from nearly 72,000 causes of impairment.[5] Nationwide, more than 50% of all impairments are caused by nutrients and sediment, metals including mercury, and pathogens (e.g., fecal coliform, *E. coli*, enterococcus bacteria, and other). Pathogens are the largest single cause of impairments.[6]

Development of TMDLs does not equate automatically with restoration that attains water quality standards. Implementation is complex and highly dependent on state and local stakeholders. It takes time, as does demonstrating results. EPA has adopted a goal of having 3,360 waters (8.2% of all impaired waters identified in 2002) in attainment with applicable water quality standards by 2015.[7]

Revising the TMDL Rules

In 1999, EPA proposed revisions to the 1992 TMDL regulations to clarify and strengthen the program. The key proposed changes included a new requirement for a more comprehensive list of impaired and threatened waterbodies; a new requirement that states, territories and authorized Indian tribes establish and submit schedules for establishing TMDLs; a new requirement that the listing methodologies be more specific, subject to public review, and submitted to EPA; clarification that TMDLs include 10 specific elements; a new requirement for a TMDL implementation plan (section 303(d) does not mandate implementation, a widely acknowledged gap; see "Factors Affecting Implementation," below); and new public participation requirements.

EPA's proposal had few strong supporters, for varying reasons. States, which would be directly affected by the proposal, criticized the burdens that new requirements would place on them. They were concerned that they lack the resources to meet tight deadlines for developing and implementing TMDLs. Further, states said that TMDLs should not necessarily be prioritized over and should be integrated with other elements of existing water quality management programs. Industry groups were greatly concerned about impacts of new pollution control requirements that result from TMDLs. But, municipal and industrial point source groups urged states and EPA to ensure that TMDL requirements do not fall disproportionately on their discharges, while possibly failing to address nonpoint source contributions to impaired waters. Farm groups and others with nonpoint discharges questioned EPA's authority to include nonpoint source pollution in the TMDL program. The forestry industry vigorously criticized potential impacts of the proposal. Environmentalists, who support the need for a stronger and more comprehensive TMDL program, objected to the lengthy time periods in the proposal before water quality improvements are likely to occur. They have criticized the lack of aggressive implementation of a program that has existed in the law since 1972.

Congressional interest was high: by the time the final rule was signed in July 2000, 13 congressional hearings had been held, and a number of legislative proposals to modify the Clean Water Act or delay the rule had been introduced.[8] EPA attempted to respond to the criticism with flexibility on some of the most contentious points. While the revised rule was undergoing final review, Congress adopted a provision in the FY2001 Military Constructions/FY2000 Urgent Supplemental Appropriations Bill (P.L. 106-246), stating that no funds could be used in FY2000 or FY2001 to "make a final determination on or implement any new rule relative to" the 1999 proposal. Because President Clinton intended to sign the bill into law but opposed the TMDL limitation, the Administration accelerated its review, allowing the EPA Administrator to sign it before President Clinton signed the appropriations bill. In the final rule, EPA acknowledged Congress's action in the legislation and delayed the effective date of the rule's changes until October 31, 2001. The text of the final rule was published on July 13, 2000.[9]

The final rule built on the existing regulatory program and added details, specific requirements, and deadlines requiring states to implement plans to clean up polluted waters. It retained key elements of the 1999 proposal for more comprehensive identification of impaired waters, schedules and minimum elements for TMDLs, and new public participation requirements. For some interested parties, what was most of interest was what was not included in the final rule. In efforts to respond to criticism, EPA dropped several provisions that were most controversial, including some potentially affecting agriculture and forestry.[10]

The Bush Administration announced in October 2001 that it would delay the effective date of the rule until May 2003, to allow for further review, but in March 2003, EPA withdrew the 2000 rule in order to consider initiating an entirely new rule or other options. Officials said that additional time was needed to decide whether and how to revise the program and that allowing the rule to take effect in May 2003 would disrupt the ongoing review. No further timetable was announced.

One EPA view, widely reported at the time, was that a new rule is not essential, because states are improving and will continue to improve the pace at which TMDLs are established, even under existing rules. Many environmentalists have said that, short of retaining the 2000 rule, the best action would be to leave the 1992 rules in place, because, although flawed, those rules are preferable to a new rule that might weaken the program. Other stakeholders have urged EPA to adopt different strategies. Many states and industries favor a rule with more flexibility than either the 2000 rule or existing regulations. In mid-2002, EPA developed a draft revised rule which it informally circulated among interest groups and federal agencies for many months, but it did not propose a new rule. In early 2009, there were some reports of discussion among EPA officials of reviving review of the 2000 rule, but the Obama Administration has not come forward with proposed regulatory changes.

As a result, program requirements under the 1992 regulations and court-sanctioned TMDL schedules remain in place and are the drivers for much of the current TMDL activity.

Recent Developments and Challenges

The earliest TMDLs developed by EPA and states often addressed impairments resulting from single pollutants and a limited number of sources. Developing even these relatively straightforward TMDLs presented substantial technical and scientific challenges, but the national program has by now become a core element of overall efforts to protect and restore water quality. States and EPA develop several thousand TMDLs annually (e.g., EPA approved 2,566 in 2010 and 2,820 in 2011). At the same time, the 303(d) program has evolved, and especially during the last decade, EPA and states have addressed more complex issues, including TMDLs for less-traditional causes of impairment, TMDLs involving both point and nonpoint sources, and multi-jurisdictional TMDLs. Legal challenges continue to influence implementation of the program.

Administrative Developments

Several provisions of the CWA direct states to report periodically to EPA on water quality conditions and trends. The 303(d) requirement to identify impaired waters is one of these provisions. Others are found in CWA section 305(b), which calls for a biennial assessment of all navigable waters in the state, and section 314, which requires a biennial report on water quality conditions of lakes. Since 2001, EPA has taken steps to minimize the administrative burden on the states of implementing these overlapping provisions by integrating the requirements into a single submission that enables a broad-scale, national inventory of water quality conditions. These Integrated Report (IR) requirements are detailed in guidance memoranda; the most recent was issued in March 2011.[11] The IR format uses a five-part categorization approach for classifying the status for each water segment, ranging from Category 1 (all designated uses are supported, and no use is threatened) to Category 5 (at least one designated use is not supported or is threatened, and a TMDL is needed—these are the waters that comprise 303(d) lists).

EPA has issued a number of guidance and other documents to support states as they address emerging TMDL issues and concerns, which are discussed next.

Addressing Nonpoint Sources in TMDLs

The traditional focus of the CWA has been on controlling direct pollutant discharges to surface waters from municipal and industrial point sources, through regulations and permits. Over time, as these sources have abated pollution, uncontrolled nonpoint sources have become a larger relative portion of remaining water quality problems. Nonpoint source pollution results from indirect discharges to surface waters from diffuse sources such as land runoff, precipitation, atmospheric deposition, drainage, seepage, or hydrologic modification. Nonpoint sources (both urban and rural) are known to cause or contribute to water quality impairments throughout the United States.

Whether and how to address nonpoint sources has been a key issue in the TMDL program. As previously described, when states or EPA develop a TMDL, they have flexibility in making use of available water quality management tools to allocate pollutant reductions. They can, for example, tighten discharge limits on point sources through enforceable permits. But municipalities and industries contend that limiting TMDLs only to point sources imposes disproportionate requirements on their operations, especially in waters that are impaired both by point and nonpoint sources.

Farming and forestry groups have long been concerned about how their activities might be addressed in TMDLs and whether they might be subject to CWA regulation of some sort, even though the act does not provide EPA with regulatory authority over nonpoint sources. EPA only can influence activities of nonpoint sources through use of grants and funding, such as CWA section 319, which addresses nonpoint source pollution through state-run nonpoint pollution management programs.[12] CWA section 303(d) does not specify whether TMDLs should cover nonpoint sources, but EPA's long-standing interpretation has been that nonpoint sources of polluted runoff should be addressed, along with point sources, where they contribute to water quality impairment. That interpretation was upheld in a key court case in 2002.[13]

Because their pollutant contributions are intermittent and because the principal policy tools available to government are voluntary or incentive-based programs, not regulatory programs, addressing nonpoint pollution sources in TMDLs is more challenging than is addressing point sources. A TMDL must include a "reasonable assurance" component that needed load reductions can be attained, but making this demonstration can be problematic when nonpoint sources are a significant source of impairment, since there is no permit or similar enforcement mechanism to provide accountability. Tracking implementation of practices to manage nonpoint pollution and verifying results are difficult, especially practices that are done voluntarily and without federal or state cost-share assistance. Some state officials contend that TDMLs for restoring impaired waters are ineffective when nonpoint sources with few or no controls are the main sources of impairment.[14] One observer, however, has suggested that states avoid allocating reductions to point sources "by relying on rather fanciful reductions from nonpoint dischargers (which, because they have no permits, can be as fanciful as one wishes)."[15]

Stormwater

Throughout the United States, stormwater discharges are responsible for thousands of water quality impairments. Stormwater is generated when precipitation from rain and snowmelt events flows over land or impervious surfaces (paved streets, parking lots, and building rooftops) and does not percolate into the ground. It may be conveyed through pipes or ditches, or it may flow over land and discharge directly into a waterbody. As the runoff flows over the land or impervious surfaces, it accumulates debris, chemicals, sediment, or other pollutants that could harm water quality if the runoff is discharged without treatment. The most common pollutants coming from stormwater are nutrients, pathogens, sediment, and metals, according to EPA.

Most stormwater sources are point sources, which are regulated under the federal stormwater regulatory program and thus are subject to CWA permit requirements. This program applies to stormwater discharges from several broad categories: municipal separate storm sewer systems (MS4s) that serve populations greater than 100,000 and smaller MS4s in Census-defined urbanized areas; construction activity one acre and larger; and specified categories of industrial activity.[16]

Other stormwater sources are nonpoint sources and thus are nonregulated by the CWA. Sources include runoff from residential and commercial application of fertilizer and manure, land disturbance and erosion, and atmospheric deposition.

Many impairments result from point source and nonpoint source stormwater discharges, and TMDLs can address both. In developing these TMDLs, data needed to characterize loads from regulated stormwater activities are generally available from CWA permitting programs. Assigning wasteload allocations to these sources is done by adjusting their permit requirements. Data to characterize loads from unregulated stormwater sources can be more difficult to obtain, and assigning cleanup obligations to them presents the types of allocation and implementation challenges previously discussed concerning nonpoint sources generally. The TMDL also must recognize that stormwater may not be the only source of pollutants such as nutrients or metals in an impaired waterbody.[17]

Mercury

Mercury is the cause of more than 4,700 impairments in thousands of 303(d) listed waterbodies; more than 60% of these impairments are indicated by high mercury levels in fish. Mercury accumulates in fish tissue as methylmercury, the form that presents the greatest risk to human health through consumption of contaminated fish.

In many waterbodies, mercury originates largely from air sources, such as coal-fired power plants, municipal waste combustors, and medical waste incinerators that deposit mercury in waters or on adjacent lands that then wash into nearby waters. Contributions may come from a combination of local, regional, and international sources. In some cases, the presence of mercury results from past practices, such as legacy gold mining, or from geologic deposits. Some mercury may be discharged to water from existing industrial point sources, although in many waterbodies, the amounts are very small compared to other sources, according to EPA.

Since 1996, states and EPA have developed nearly 7,000 mercury TMDLs. Given the variety of potential mercury sources, developing and implementing TMDLs for mercury-impaired waters often involves coordination among multiple programs—water, air, waste, and toxics programs— and multiple stakeholders. Technical and organizational challenges can be significant, particularly where mercury is predominantly from atmospheric sources. Water management programs generally do not control sources of air emissions, and controlling deposition that originates outside a single jurisdiction is difficult.

Two TMDLs that notably have addressed these types of challenges are a statewide mercury TMDL developed in Minnesota, consisting of two regional TMDLs in the state that cover 998 waterbody/pollutant combinations, and a regional mercury TMDL developed by seven northeast states (Connecticut, Maine, Massachusetts, New York, New Hampshire, Vermont, and Rhode Island).[18] These and many other mercury TMDLs expect to be accomplished over time through existing and future regulatory controls outside of the CWA (such as Clean Air Act regulation of mercury emissions from coal-fired powerplants and other sources), making it difficult to estimate with confidence the effectiveness of mercury reduction strategies. Thus, most anticipate periodic reevaluation by the seven participating states to determine progress and the need for additional or revised reductions to meet water quality standards.

Ocean Acidification

Oceans naturally absorb carbon dioxide (CO_2) from the atmosphere, but too much CO_2 can cause them to become overly acidic. Scientists believe that ocean acidification (OA) is primarily caused by increasing CO_2 concentrations in the atmosphere. In a 2010 report, the National Research Council of the National Academies of Sciences concluded that ocean chemistry is changing at an unprecedented rate and magnitude due to anthropogenic CO_2 emissions and that "there is a risk of ecosystem changes that threaten coral reefs, fisheries, protected species, and other natural resources of value to society."[19]

Activists have been pressing EPA on a variety of fronts to address the role of increasing emissions of CO_2, a greenhouse gas, on climate change, including under the TMDL program. In 2009, the Center for Biological Diversity (CBD) sued EPA in a challenge to the agency's approval of Washington's 303(d) list, because the state's list failed to include coastal waters as impaired for marine pH due to CO_2.[20] CBD also sought to force EPA to revise and strengthen its existing water quality criterion for pH in marine waters, issued in 1976. Under the CWA, EPA issues water quality criteria, which are scientific information regarding concentrations of specific chemicals in water which protect aquatic life or human health. States, in turn, use these EPA-recommended criteria as a basis for developing water quality standards. CBD argued to EPA that the existing marine pH criterion is not protective enough to address decreasing pH levels due to increased CO_2 emissions. The group contended that if states adopted more stringent marine pH criteria in their water quality standards, more waters likely would be identified as impaired, thus requiring TMDLs to achieve reduction in CO_2 emissions that are responsible for the standards violations. EPA has said that it has insufficient data to revise the national marine pH criterion at this time.

To settle the CBD lawsuit, EPA agreed to look for opportunities to use the CWA to address OA (it did not commit to revising the existing marine pH water quality criterion). In 2010, EPA published a Federal Register Notice seeking public comment on what considerations it should take into account when deciding how to address the listing of waters as threatened or impaired by OA under the TMDL program, including how to develop TMDLs for such listed waters.[21] Also as part of the settlement, EPA subsequently issued a memorandum that advised coastal states to list waters that do not meet standards for marine pH when data are available, but the memo did not elevate in priority the assessment and listing of waters for OA. The memo said, "EPA recognizes that information is absent or limited for OA parameters and impacts at this point in time and, therefore, listing for OA may be absent or limited in many States."[22] The agency expects to provide additional guidance to states when future OA research provides improved monitoring and assessment methods.

Climate Change

Even for relatively straightforward impairments (e.g., a single pollutant such as a heavy metal discharged from known sources), TMDL development is complex and resource-intensive. The process requires extensive data on current water quality conditions, pollutants and sources causing impairments, and modeling to analyze effects of load and wasteload allocations that will attain water quality standards with a margin of safety. Climate change considerations add even greater complexity to the analyses, including the need to assess future conditions that are themselves uncertain and sources that can be both domestic and global. Nevertheless, a 2011 EPA draft report notes that climate change may alter attainability of some designated uses and parameters related to water quality standards (e.g., lower streamflow may increase stream temperature) and recommends that "TMDLs and water quality standards should be examined to ensure that these remain protective of aquatic life uses under changing climatic conditions."[23]

The number of future TMDLs that could or should include potential climate change effects is unknown. According to EPA, "It is probable that most existing TMDLs do not take climate change considerations into account, and due to the number of TMDLs in existence (~40,000) it is not feasible to re-open each TMDL with the sole purpose of incorporating climate change considerations." In the future, however, EPA will encourage development of TMDLs that incorporate projected climate impacts and uncertainty, as appropriate.[24]

Consideration of climate change currently is occurring in a few TMDLs, including Vermont's phosphorus TMDL for Lake Champlain. A TMDL is required for Lake Champlain because phosphorus concentrations in many segments of the lake are higher than levels allowed in Vermont water quality standards. The states of Vermont and New York jointly developed a phosphorus TMDL, which EPA approved in 2002. However, EPA's approval of the TMDL was challenged in a federal lawsuit, which asserted a variety of flaws, including insufficiently stringent wasteload allocations of pollutants for point sources coupled with a lack of reasonable assurances that nonpoint source reductions would occur; an inadequate margin of safety; inadequate specificity of the stormwater component of the wasteload allocations; and failure to consider water resource effects associated with documented and predicted climate change.[25]

In April 2010, EPA signed a settlement agreement with the environmental plaintiffs, and in January 2011, the agency reconsidered its previous approval of the Lake Champlain TMDL and disapproved the Vermont portion. After re-reviewing the 2002 TMDL, EPA concluded that two of the four contested elements are not consistent with EPA regulations and guidance available at the time of TMDL approval: margin of safety, and reasonable assurance. EPA found the two other elements of the complaint—the stormwater component, and climate change considerations—to be consistent with then-current rules and guidance.

Under section 303(d)(2) of the CWA, upon disapproval of a TMDL, EPA must establish a new TMDL to implement applicable water quality standards, and EPA has begun a new phosphorus TMDL for the impaired Vermont segments in Lake Champlain. EPA informed the state of Vermont that, in doing so, it may refine several aspects of the TMDL, not just the components that were determined to be inadequate. Although EPA did not reject the challenged climate change considerations in the 2002 TMDL, apparently they are being addressed now: one component of preparing the TMDL is a study of potential effects of climate change and phosphorus loads to Lake Champlain. EPA expects to prepare the TMDL for public comment in late 2012 or early 2013.[26]

Multi-Jurisdiction TMDLs

For some time, EPA has encouraged states to develop TMDLs using a watershed approach, recognizing that, where multiple impaired segments are clustered within a watershed, it frequently is more efficient to organize restoration activities across the watershed, instead of doing TMDLs for individual stream segments. One of the first states where this approach was used is Montana, but it is also reflected in other locations. In cases where the watershed spans more than one state, TMDLs have been developed jointly or in coordination by several states. Beyond the expected challenges of TMDL development in a single stream segment (e.g., data, resources, scientific uncertainties), these multi-jurisdiction TMDLs present additional challenges (e.g., differing priorities, institutional coordination).

- Several states have developed statewide TMDLs for a single pollutant, such as Minnesota's mercury TMDL (described above). Other states (e.g., Florida and North Carolina) are developing statewide TMDLs for mercury-impaired waters. Several states also have developed statewide bacteria TMDLs (Rhode Island, Connecticut, New Hampshire, and Vermont). New Hampshire has developed a statewide chloride TMDL.
- Because air deposition from outside a single region is a prominent source of mercury impairments, many mercury TMDLs are being developed on a multi-jurisdiction basis, including the six-state Northeast Regional Mercury TMDL discussed above.
- New York and Connecticut developed TMDLs with measures needed to attain water quality standards for dissolved oxygen in Long Island Sound. The TMDLs require wastewater treatment plants and other point sources in both states to reduce nitrogen discharges that contribute to the Sound's water quality impairment.
- Water quality monitoring data show that approximately half of the length of the Ohio River (475 miles) is now listed as impaired for contact recreational use. As a result, a bacteria TMDL is now under development for the entire river due to the diverse sources that contribute to impairments, including municipal sewer overflows, animal feedlots, textile and pulp mills, as well as nonpoint sources such as leaking septic systems and land application of manure. Two EPA regions, the six states along the river, and the Ohio River Valley Water Sanitation Commission (ORSANCO) are participating in the TMDL.
- In 2010, EPA approved a TMDL developed by the Washington Department of Ecology to address nutrient enrichment problems which have caused algae blooms and low dissolved oxygen levels in Lake Spokane. The TMDL applies to a 63-mile stretch of the Spokane River from the Idaho-Washington border to the lake. It is one of the first TMDLs to coordinate point source reductions across state boundaries; it includes cross-boundary wastewater and stormwater sources in Idaho, because the Spokane River originates in that state, and modeling determined that those point sources contribute to water quality impairments in the watershed in Washington. Among the challenges for Washington and EPA was developing a TMDL with equitable allocations for dischargers in Washington and Idaho, one

that would not call for disproportionate pollutant reductions in either state.[27]

The largest multi-jurisdictional TMDL, for the Chesapeake Bay watershed, is discussed next.

Chesapeake Bay TMDL

One TMDL that touches on several of the issues discussed above and that has drawn considerable attention and controversy, including from Members of Congress, was developed by EPA for the 64,000-square-mile Chesapeake Bay watershed.

Despite several decades of activity by governments, the private sector, and the general public, efforts to improve and protect the Chesapeake Bay watershed have been insufficient to meet restoration goals. Although some specific indicators of Bay health have improved slightly or remained steady recently (such as blue crabs and underwater bay grasses), others remain at low levels of improvement, especially water quality. Overall, the Bay and its tributaries remain in poor health, with polluted water, reduced populations of fish and shellfish, and degraded habitat and resources. The primary pollutants causing impairments are nutrients (nitrogen and phosphorus) and sediment discharged from multiple urban, suburban, and rural sources around the Bay.

In May 2009, President Obama issued an executive order that declared the Bay a "national treasure" and charged the federal government with assuming a strong leadership role in restoring the Bay.[28] The executive order established a Federal Leadership Committee for the Chesapeake Bay to develop and implement a new strategy for protecting and restoring the Chesapeake region. The resulting strategy, which was released in May 2010, launched major specific environmental initiatives to establish new clean water regulations on stormwater discharges and pollution discharges from animal feedlots in the Bay watershed, put new agricultural conservation practices on farms in the region, and restore land and water habitat.[29]

A central feature of the overall strategy is EPA's establishment of a TMDL for Chesapeake Bay, which was necessitated because the Bay states' previous restoration efforts were insufficient to attain water quality standards. Under a consent decree resolving litigation over impairment of Bay waters, EPA was required to establish a Chesapeake Bay TMDL no later than May 1, 2011.[30] EPA issued the TMDL on December 29, 2010.

The Chesapeake Bay TMDL—which is actually divided into three separate TMDLs—is the largest single TMDL developed to date. It addresses all segments of the Bay and its tidal tributaries that are impaired from discharges of nitrogen, phosphorus, and sediment, with a goal of having TMDL implementation measures in place by 2025. Nitrogen and phosphorus are considered the main contributors to poor Bay water quality because, in excess amounts, they spur algae blooms, which block sunlight critical to underwater grasses that support crabs, fish, and waterfowl. When the algae die, they sink to the bottom and decompose in a process that depletes the water of oxygen, creating so-called dead zones. Sediment also depletes water of oxygen. To meet the TMDL goals, the amount of nitrogen entering the Bay needs to be reduced by 75.39 million pounds a year, phosphorus needs to be cut by 14.55 million pounds, and sediment needs to be reduced by 3.7 million tons from 2009 levels. The TMDL allocates needed reductions of these pollutants to all jurisdictions in the watershed. Detailed plans identifying specific reductions are to be developed by the seven jurisdictions located in the Chesapeake Bay watershed in Watershed Implementation Plans (WIPs).[31]

As part of the TMDL process, states are to prepare WIPs in three phases identifying specific reductions and control measures to achieve needed pollutant reductions from point sources (i.e., industrial and municipal facilities and large animal feeding operations) and nonpoint sources (i.e., farms and forests). The WIPs are part of the accountability framework for the Bay TMDL. Phase I WIPs were developed in December 2010. They generally outlined the types of controls and best management practices (BMPs) that will be utilized to meet the first major goals of the TMDL: that 60% of needed practices to achieve water quality standards will be in place by 2017. Each of the states' plans is different and reflects the states' choices regarding localized identification of needed controls and BMPs. EPA is now reviewing the Phase II WIPs, which describe how the jurisdictions will work with specific localities over the next five years to reduce sources of impairment. The Bay jurisdictions are to submit Phase III WIPs by 2017, including any modifications needed to meet the 2025 goal of having in place 100% of practices needed to achieve the nutrient reduction targets.

The use of WIPs as part of the restoration strategy provides states flexibility to determine the mix of specific controls they deem appropriate to meet the overall reduction goals. EPA fully expects that the jurisdictions will meet commitments in the WIPs, but the agency also identified a number of potential "backstop" actions available to it under the CWA if a state fails to do so, including expanding permit coverage to currently unregulated sources, requiring net improvement offsets, conditioning EPA grants, or increasing federal enforcement in the watershed. These CWA backstop authorities are always available to EPA, but EPA's use of them in the Chesapeake Bay TMDL has been very controversial. It also marks a change from the historic nature of Bay restoration, which was based primarily on stakeholder agreements.[32] EPA's review of the WIPs has led to discussions with each jurisdiction to resolve certain issues, such as providing reasonable assurance that allocation targets will be achieved. EPA's approval of the Phase I WIPs included certain backstops for New York, Pennsylvania, and West Virginia.[33]

The Chesapeake Bay TMDL has a number of novel elements. For example, building on a 2009 voluntary agreement by the Bay states, the TMDL requires the jurisdictions to commit to and report on short-term milestones every two years to help government and the public assess progress in implementing cleanup plans. Federal and state officials evaluated progress under the first two-year milestone goals, which were to be reached by the end of 2011, and concluded that in general, states have made significant overall progress and accelerated nutrient reduction efforts, but that improvements are still needed in many areas. The mixed findings were not unexpected. A review by environmental groups for a subset of important pollution controls found that each state was ahead of schedule in implementing some actions, such as nutrient reduction from wastewater treatment plants, but behind in others, such as implementing agriculture conservation management plans.[34] The first milestones under the TMDL were submitted by states in January 2012, and they must be attained by the end of 2013.

The TMDL also reflects the view that nutrient trading can be a promising strategy for meeting nutrient load limits in a cost-effective way. Under this market-based strategy, sources such as municipal wastewater and industrial facilities achieve their individual load limits by purchasing load reductions from other sources such as farmers for a lower cost than if the point source were to install technology on-site. Three states in the Bay watershed (and a number of other states outside the region) have initiated nutrient trading programs, but little implementation has occurred. In the Bay TMDL, EPA supports (but does not mandate) trading to meet the plan's nutrient reduction targets and to account for and manage new or increased loadings in the future, and each of the Bay jurisdictions included nutrient trading in its Phase II WIP.

Many supporters believe that trading has the potential to result in substantial overall cost savings, but actual predictions depend on numerous considerations and uncertainty (e.g., required trading ratios, potential tradeable loads, geographic scope of trading, transaction costs). Trading is also seen as a means of encouraging agriculture's participation in Bay restoration, since it has the potential of generating new revenue for the agricultural sector (since a point source may pay a farmer to reduce pollutants that the point source would otherwise have to control). It is generally believed that installing measures to control agricultural sources of nutrients is more cost-effective than, for example, urban nonpoint source practices such as street sweeping or installing urban forest buffers.[35] To support these types of efforts, USDA awarded $2.35 million to five projects in FY2012 to promote water quality trading in the Chesapeake Bay watershed. The supported projects are intended to help completion of state water quality market rules and organizations needed to carry out trading between point sources and farmers and ranchers. At the same time, some observers are cautious about the role of water quality trading in restoring impaired waters. One recent assessment of nutrient trading in the Bay watershed summarized both positive aspects of nutrient trading—such as potential to use market forces and increase stakeholder support—and negative aspects—trading to reduce water pollutants is technically complicated, interstate trading is hindered by differing state requirements, and verifying results is difficult.[36]

EPA's TMDL and the overall federal Bay restoration strategy under the executive order are controversial with a number of groups that are concerned about the likely mandatory nature of many of EPA's and states' upcoming actions. A major concern is that the strategy will be very costly to implement and will hamper the region's economic recovery. Critics point out that EPA has not prepared a cost-benefit analysis of restoring the watershed. EPA is now preparing such an analysis, based on the Phase I WIPs, and expects to complete it by the end of 2012. Estimating benefits is more complicated than estimating costs because of the difficulty of monitoring and verifying results, according to EPA.

Implicit in much of the criticism of the Chesapeake Bay TMDL is concern that EPA will use it as a model for developing similar large-scale TMDLs in other locations. As previously described, EPA has developed and approved a number of multi-state TMDLs (e.g., the Northeast Regional Mercury TMDL), but some observers worry that what they view as an aggressive and inflexible TMDL strategy for the Chesapeake Bay will be replicated for other waterbodies.

Legal challenges were brought by agricultural and home builder groups, who argue that EPA has exceeded its CWA authority (for example, by imposing detailed pollutant allocations and using backup authority to strengthen WIPs) and impinged on the responsibilities of states to assign pollution reductions to individual sources.[37] They also question the science and accuracy of computer models that EPA used in setting overall pollution limits. EPA contends that development of the TMDL was well within its authority under the CWA, and the agency disputes that it overrode state authority, noting that states were fully involved in the TMDL development process. Several environmental groups and municipal water utility groups have intervened on EPA's side in the litigation.

Environmental activists have expressed support that the federal government is finally asserting a leadership role to restore the Bay and have endorsed legislation that would codify requirements of the Bay TMDL in the Clean Water Act, while authorizing grants and other assistance for implementing required measures. Companion bills to do so were introduced in the 111th Congress (S. 1816 and H.R. 3852), but no legislation was enacted. The House Agriculture Committee also approved separate legislation in the 111th Congress (H.R. 5509) that would have authorized an expanded role for the Department of Agriculture in Chesapeake Bay restoration, to counter EPA's role in the TMDL. Legislation similar to H.R. 5509 has been introduced in the 112th Congress (H.R. 4153). Congress has shown interest in the impact of the plan on agriculture through oversight hearings by the House Agriculture Subcommittee on Conservation, Energy, and Forestry (on March 16 and November 3, 2011).

Outside Assessments of the TMDL Program

Over the last decade, several assessments have recommended ways to strengthen the TMDL program. One Government Accountability Office (GAO) report examined whether available water quality data are sufficient to allow state officials to make key decisions about activities such as identifying waters that do not meet water quality standards and developing strategies to address those waters. Water quality data are so limited, particularly data for nonpoint sources, that many fear that TMDL decisions will be based on unsound information and will impose unneeded or inappropriate control mandates, according to GAO. Inconsistent monitoring, data collection, and listing procedures used by states to identify impaired waters have hindered efforts to develop effective TMDL programs, GAO found.[38]

In 2002, GAO issued a related report that examined the different approaches used by states to identify impaired waters. Some of the approaches have no appropriate scientific basis, GAO said, and states apply a range of quality assurance procedures to ensure the quality of data used to make impairment decisions. GAO concluded that, because of inconsistencies in states' approaches to identifying impaired waters, the information in EPA's database of impaired waters is of questionable reliability, and EPA cannot reliably tally the number of TMDLs that must be completed nationwide.[39]

EPA's responses to both reports are reflected in the Integrated Report guidance in place since 2004 (see "Administrative Developments"). It provides guidance on how states can document the scientific and technical rationale for categorizing their waters, and it lists key elements that must be included in states' descriptions of their methodologies. In addition, in order to encourage a long-term process of incremental improvement in state water monitoring programs, EPA published a document recommending the basic elements of such a program.[40]

Following EPA's controversial revisions of the TMDL program regulations in 2000 (which were subsequently withdrawn in 2003), the agency's FY2001 appropriation bill, P.L. 106-377, required a study by the National Academies of Sciences (NAS) on the scientific basis of the program. The NAS report, issued in 2001, concluded that scientific knowledge exists to move forward with the program but recommended certain types of changes, such as utilizing iterative, adaptive implementation and revision, as scientific data and information improve.[41] EPA's review of the 2000 rule after 2001 was at least partly to consider how to respond to those recommendations, and many of the more recent TMDLs explicitly allow for and anticipate use of adaptive management techniques.

In 2007, EPA's Inspector General (IG) reported on TMDL implementation, particularly EPA's attention to information and systems to track and evaluate implementation. The IG found that a lack of information prevents EPA from determining if TMDL implementation activities are occurring in a timely manner, and the extent to which TMDLs are restoring impaired waters. EPA tracks specific outcomes—e.g., numbers of TMDLs developed—but not functional results of water quality improvement. The report noted that EPA has recently taken steps to measure results and improve program data, including sponsoring studies of TMDL implementation and evaluating additional results measures. These tasks are challenging, the IG acknowledged, but the report urged EPA to provide more management direction to improve its ability to assess how well the program is functioning.

EPA concurred with many of the IG's conclusions, but responded in part that the agency has only limited authority to require new post-TMDL monitoring, data tracking, and reporting. The IG had recommended that EPA report annually on nationwide implementation activities, such as BMPs completed and inclusion of wasteload allocations in point source discharge permits. EPA said that "there are substantial cost and logistical implications of reporting annually and nationally on TMDL implementation and TMDL environmental outcomes. After a point, tradeoffs between level of effort invested in reporting and actually restoring impaired waters must be made by EPA and the States." [42]

Factors Affecting Implementation

EPA and other stakeholders frequently point out that TMDLs are not the only approach to restoring water quality and that TMDLs are sometimes inseparable from the combined effect of multiple surface water protection program activities. TMDL implementation is only one of several factors contributing to water quality improvements, and it is likely to be a small percentage, the agency says. Other factors besides TMDL implementation could influence whether a waterbody is in attainment with water quality standards, including new monitoring data or a change in water quality standards, with data showing that the waterbody meets the new standards.

Further, states point out that there can be advantages and disadvantages of using TMDLs to address impairments. Some say, for example, that relying on TMDLs is the right approach when the problem and the source of impairment are unknown, and more scientific evidence is needed to allocate pollutant reductions. Others say that doing TMDLs for some types of sources (e.g., agriculture) provides transparency and assurance to the public about what the state is doing to improve water quality. But in view of the technical and resource challenges of doing TMDLs, in some cases, alternative approaches may also be appropriate. For example, a state may lawfully avoid listing an impaired water as needing a TMDL if it can demonstrate to EPA that other pollution control programs provide reasonable assurance that water quality standards will be met, such as imposition of point source permit limits, or state regulations and local ordinances for nonpoint sources. However, this use of non-TMDL options is a relatively new idea and has been little used. Some think that alternative approaches are more difficult to achieve than a TMDL. One approach, in the state of Washington, is called "straight to implementation." It involves working directly with landowners, which enables actions to prevent pollution without first developing a TMDL, especially when the state has clear enforcement authority over all types of sources, including nonpoint. Also, a TMDL may not be necessary in cases where the source of impairment is clearly known and control measures and funding are known and available.[43]

That said, when a TMDL is developed, implementation is a huge gap in the program for several reasons. First, as noted previously, CWA section 303(d) requires development of a TMDL for impaired waters but does not require implementation. States' strategies for implementation vary widely. A recent survey determined that five states have laws requiring TMDL implementation plans, either separately after the TMDL is approved or during development of the TMDL. In five other states, formal implementation plans are not required, but are completed pursuant to state guidance. In a third category are 15 states that develop implementation plans based on an unwritten state initiative or they rely on federal guidance. And finally, another 15 states have less intensively developed implementation plans, which generally are a brief section in the TMDL report. No information was found for nine states.

The survey also examined how states implement nonpoint source pollutant reductions. It found that 10 states have regulations to address some or all types of nonpoint source pollution, while most states use incentive-based and voluntary programs instead of a regulatory approach.[44]

Second, a number of barriers to TMDL implementation can be identified.[45]

The most prominent hindrance is funding. The CWA provides no dedicated funding for TMDL development or for implementation. Inadequate resources are cited as a problem for BMP implementation for nonpoint sources, for point source upgrades such as wastewater treatment plants, for abandoned mine land restoration, for staffing and technical expertise, and for data collection, modeling, and follow-up monitoring. Technical impediments are a related challenge, such as insufficient scientific data on the pollutant removal performance of BMPs for nonpoint sources, or inability to link water quality impacts to sources or causes due to lack of data and understanding of pollutant fate and transport.

At the same time, several factors that seem to aid effective implementation also can be identified. These include active involvement of stakeholders and local and state governments; using a diversity of approaches to address sources; existence of a watershed plan that is focused on issues in the watershed and achievable through corrective actions; targeted or phased implementation; and adequate resources to implement corrective measures, monitor progress, and provide technical assistance.

Conclusion

After nearly 40 years of implementing the CWA, EPA and states acknowledge that a substantial portion of the nation's waters still are impaired or threatened by pollution. The most recent national inventory of water quality reported that nearly 40% of surveyed water bodies remain too polluted for fishing, swimming, and other designated uses. Yet those numbers only represent rivers, streams, and lakes actually surveyed by state monitoring programs—typically about one-third of all waters. The TMDL assessments developed by states yield more precise water quality information and identify large numbers of waters requiring additional measures before water quality standards are attained. The TMDL program is in a period of transition. Many states are emerging from earlier consent decree mandates and are increasingly addressing new challenges—for example, more complex and resource-intensive TMDLs, larger scale impairments, and nonpoint sources. Whether the program as it now exists is well suited to address some of these problems, such as ocean acidification or climate change, is debatable. In August 2011, EPA and state program managers launched discussions of developing new goals for the program. One year later, these discussions produced a draft "long-term vision" for reforming the process, including allowing states the option to consider protecting healthy waters, using alternative approaches that incorporate adaptive management, and integrating TMDLs with other CWA and Safe Drinking Water Act programs.[46]

Other than recent oversight hearings on the Chesapeake Bay TMDL, Congress has not shown active interest in the TMDL program for more than a decade. Some stakeholders, especially states, believe that several issues present Congress with an opportunity to examine the TMDL provisions of the CWA. Issues could include integrating TMDLs into a larger clean water program that considers all steps—from designation of uses to implementation—in order to meet water quality standards, recognizing and striking a balance between water quality restoration and pollution prevention, changing focus from point sources to nonpoint sources, and addressing resource and funding needs.

End Notes

[1] Reference throughout this report to states includes states, U.S. territories, and Tribes authorized by EPA to administer water quality standards on tribal lands.

[2] U.S. Environmental Protection Agency, National Water Quality Inventory: Report to Congress, 2004 Reporting Cycle, EPA 841-R-08-001, January 2009, http://water.epa.gov/lawsregs/guidance/cwa/305b/2004report_index.cfm.

[3] For information on TMDL litigation by state, see http://water.epa.gov/lawsregs/ lawsguidance /cwa/tmdl/lawsuit.cfm. However, this information is a partial list that is not comprehensive or current.

[4] This is a longer time frame than has been mandated as a result of some of the TMDL litigation. The schedules for TMDLs in lawsuits concluded by consent decrees and settlement agreements range from 4 years to 20 years; most call for a 10-year development schedule.

[5] Many waters are impaired by more than one pollutant or other cause.

[6] For information, see http://ofmpub.epa.gov/tmdl_waters10/attains_nation_cy. control? p_report_type= T#status_of_data.

[7] U.S. Environmental Protection Agency, Office of Water, National Water Program Guidance, Fiscal Year 2013, April 2012, p. 32. In 2002, states identified 39,503 specific waterbodies as impaired.

[8] During the 106th Congress, hearings were held by the House Agriculture Committee; House Transportation and Infrastructure Committee; Senate Agriculture, Nutrition and Forestry Committee; and Senate Environment and Public Works Committee.

[9] U.S. Environmental Protection Agency, "Revisions to the Water Quality Planning and Management Regulation and Revisions to the National Pollutant Discharge Elimination System Program in Support of Revisions to the Water Quality Planning and Management Regulation; Final Rules," 65 Federal Register No. 135, July 13, 2000, pp. 43586- 43670.

[10] See CRS Report RL30611, EPA's Total Maximum Daily Load (TMDL) Program: Highlights of the Final Revised Rule, by Claudia Copeland.

[11] See "Information Concerning 2012 Clean Water Act Sections 303(d), 305(b), and 314 Integrated Reporting and Listing Decisions," http://water.epa.gov/lawsregs/l awsguidance/cwa/tmdl/ir_memo_2012.cfm.

[12] While the 319 program is voluntary at the federal level, states may include regulatory components in their 319 programs.

[13] Pronsolino v. Marcus, 91 F.Supp.2d 1337 (N.D.Cal 2000) aff'd, Pronsolino v. Nastri, 291 F.3d 1123 (CA9 2002).

[14] Amena H. Saiyid, "TMDLs Ineffective for Waters Impaired by Nonpoint Sources, State Officials Tell GAO," Daily Environment Report, vol. 141 (July 24, 2012), p. A-8.

[15] Oliver A. Houck, "The Clean Water Act Returns (Again): Part 1, TMDLs and the Chesapeake Bay," Environmental Law Reporter, vol. 41, no. 3 (2011), pp. 10208-10228, 10211.

[16] For information, see CRS Report 97-290, Stormwater Permits: Status of EPA's Regulatory Program, by Claudia Copeland.

[17] Pathogens, metals, nutrients, and sediments are the cause of nearly 57% of impairments in 303(d) listed waters. See http://ofmpub.epa.gov/tmdl_waters10/attains_nation_cy.control? p_report_type=T#status_of_data. How many of those impairments are from stormwater sources is unknown.

[18] U.S. Environmental Protection Agency, "Examples of Approved Mercury TMDLs," http://water.epa.gov/lawsregs/ lawsguidance/cwa/tmdl/mercury.cfm.

[19] National Research Council, National Academies of Science, Ocean Acidification: A National Strategy to Meet the Challenges of a Changing Ocean, National Academies Press, 2010, p. 15. For additional information, see CRS Report R40143, Ocean Acidification, by Eugene H. Buck and Peter Folger.

[20] In chemistry, pH (potential (of) hydrogen) is a measure of the concentration of hydrogen ions in a solution, hence, its acidity or alkalinity.

[21] U.S. Environmental Protection Agency, "Clean Water Act Section 303(d): Notice of Call for Public Comment on 303(d) Program and Ocean Acidification," 75 Federal Register 13537-13540, March 22, 2010.

[22] Denise Keehner, Director, EPA Office of Wetlands, Oceans and Watersheds, memorandum, "Integrated Reporting and Listing Decisions Related to Ocean Acidification," November 15, 2010, p. 4.

[23] U.S. Environmental Protection Agency, Global Change Research Program, National Center for Environmental Assessment, Implications of Climate Change for Bioassessment Programs and Approaches to Account for Effects, Preliminary Draft, EPA/600/R-11/036A, March 2011, p. 7-1, http://cfpub.epa.gov/ncea/cfm/recordisplay.cfm?deid= 233810. EPA expected to issue a final report in mid-2011 but has not done so.

[24] U.S. Environmental Protection Agency, National Water Program 2012 Strategy: Response to Climate Change, Public Comment Draft, March 2012, p. 51, http://water.epa.gov /scitech/climatechange/upload/ NWP_Draft_Strategy_03-27-2012.pdf.

[25] Conservation Law Foundation v. U.S. Environmental Protection Agency, No. 2:08-cv-00238wks (D. Vt. 2008).

[26] http://www.epa.gov/region1/eco/tmdl/pdfs/vt/LakeChamplainTMDLDevelpmtProcess.pdf.

[27] Michael A. Bussell, Director, Office of Water & Watersheds, U.S. EPA Region 10, letter to Kelly Susewind, Washington Department of Ecology, "Approval of the Spokane River Dissolved Oxygen TMDLs." May 10, 2010, p. 19, http://www.epa.gov/region 10/pdf/tmdl/spokane_tmdl_approval_may2010.pdf.

[28] Executive Order 13508, "Chesapeake Bay Protection and Restoration," 74 Federal Register 23099-23104, May 15, 2009.

[29] For information, see http://www.chesapeakebay.net/news_ federalstrategy.aspx? menuitem= 51207.

[30] Fowler v. U.S. EPA, Case No. 1:09-CV-00005-CKK (D.D.C.), May 10, 2010.

[31] New York, Pennsylvania, Maryland, Delaware, Virginia, West Virginia, and the District of Columbia.

[32] U.S. Government Accountability Office, Chesapeake Bay Restoration Effort Needs Common Federal and State Goals and Assessment Approach, GAO-11-802, August 2011, p. 10.

[33] In New York, EPA required installation of biological nutrient removal at all wastewater treatment plants in the Bay watershed. In Pennsylvania, EPA moved 50% of the state's stormwater allocation into wasteload allocation, meaning that the state likely will have to regulate those stormwater sources through MS4 permits. In West Virginia, EPA shifted 75% of the state's animal feeding operation (AFO) load into wasteload allocation, meaning that sources could potentially be subject to state or federal CWA permits as necessary to protect water quality if West Virginia does not achieve reductions in agricultural loads from nonpoint sources, as identified in the WIP.

[34] Chesapeake Bay Foundation and Choose Clean Water, "2011 Milestone Analysis Shows Progress," July 9, 2012, http://www.cbf.org/page.aspx?pid-3815.

[35] RTI International, Nutrient Credit Trading for the Chesapeake Bay, An Economic Study, prepared for the Chesapeake Bay Commission, May 2012.

[36] Willamette Partnership, Pinchot Institute for Conservation, and World Resources Institute, In it Together: A How-To Reference for Building Point-Nonpoint Water Quality Trading Programs, Case Studies (Part 3 of 3), July 2012, pp. 32- 44.

[37] American Farm Bureau Federation and Pennsylvania Farm Bureau v. U.S. EPA, Case No. 11-cv-0067 (M.D. Pa.).

[38] U.S. General Accounting Office, Key EPA and State Decisions Limited by Inconsistent and Incomplete Data, RCED-00-54, March 14, 2000.

[39] U.S. General Accounting Office, Inconsistent State Approaches Complicate Nation's Efforts to Identify Its Most Polluted Waters, GAO-02-186, January 11, 2002.

[40] U.S. Environmental Protection Agency, Elements of a State Water Monitoring and Assessment Program, EPA 841- B-03-003, March 2003.

[41] National Research Council of the National Academies of Science, Water Science and Technology Board, Assessing the TMDL Approach to Water Quality Management, June 2001.

[42] Office of Inspector General of the U.S. Environmental Protection Agency, Total Maximum Daily Load Program Needs Better Data and Measures to Demonstrate Environmental Results, Report No. 2007-P-00036, September 19, 2007, p. 17.

[43] See Environmental Law Institute, National Workshop to Advance State TMDL Programs, Final Project Report & Workshop Proceedings, November 2008, http://www.eli.org/pdf/tmdl/TMDL.FinalReport.pdf. Hereafter, National Workshop.

[44] Virginia Tech University, Biological Systems Engineering Department and the Center for TMDL and Watershed Studies, State-Specific TMDL Implementation Information, September 2008, http://water.epa.gov/lawsregs/ lawsguidance/cwa/tmdl/upload/2009_09_09_tmdl_results_27_st_imp_info_va_tech.pdf.

[45] See, for example, Brian Benham, Rebecca Zeckoski, and Gene Yagow, "TMDL Implementation: Lessons Learned," Proceedings: Water Environment Federation TMDL 2007 Conference, pp. 428-442, http://water.epa.gov/lawsregs/ lawsguidance/cwa/tmdl/upload/2009_09_09_tmdl_results_27_st_imp_info_va_tech.pdf; Environmental Law Institute, State TMDL Program Resource Center, "State TMDL Program Snapshots," http://www.eli.org/program_areas/Events/ StateTMDLProgramSnapshots.cfm; and National Workshop.

[46] Amena H. Saiyid, "Draft TMDL Plan Focuses on Protecting Waters, Not Just Restoring Impaired Ones," Daily Environment Report, August 20, 2012, http://www.bna.com/draft-tmdl-plan-n12884911304/.

In: Water Pollution Controls
Editor: Julia Crawford

ISBN: 978-1-62417-441-4

Chapter 3

STORMWATER PERMITS: STATUS OF EPA'S REGULATORY PROGRAM*

Claudia Copeland

SUMMARY

The Environmental Protection Agency (EPA) and states are implementing a federally mandated program for controlling stormwater discharges from industrial facilities and municipalities. Large cities and most industry sources are subject to rules issued in 1990, and EPA issued permit rules to cover smaller cities and other industrial sources and construction sites in 1999. Because of the large number of affected sources and deadline changes that led to confusion, numerous questions have arisen about this program. Impacts and costs of the program's requirements, especially on cities, are a continuing concern.

The 109th Congress enacted omnibus energy legislation (P.L. 109-58, the Energy Policy Act of 2005) that included a provision giving the oil and gas industry regulatory relief from some stormwater control requirements. In May 2008, a federal court vacated an EPA rule implementing this provision. EPA intends to issue a revised rule that repeals the one that was vacated by the court and codifies the statutory exemption in P.L. 109-58, but the agency does not have a specific schedule for doing so. In the 111th Congress, the House passed a bill that

* This is an edited, reformatted and augmented version of a Congressional Research Service publication, CRS Report for Congress 97-290, prepared for Members and Committees of Congress, from www.crs.gov, dated July 30, 2012.

included a provision that would repeal the exemption in P.L. 109-58 (H.R. 3534), but the Senate took no action.

Congress often looks to federal agencies to lead or test new policy approaches, a fact reflected in legislation enacted in the 110th Congress. Section 438 of the Energy Independence and Security Act (P.L. 110-140, EISA) requires federal agencies to implement strict stormwater runoff requirements for development or redevelopment projects involving a federal facility in order to reduce stormwater runoff and associated pollutant loadings. EPA has issued technical guidance for federal agencies to use in meeting these requirements.

In 2009 the National Research Council issued a report calling for major changes to strengthen EPA's stormwater regulatory program, which it criticized as being inconsistent nationally and failing to adequately control all sources of stormwater discharge that contribute to waterbody impairment. In response, EPA has begun efforts to expand regulations and strengthen the current program with a revised rule that it expects to propose in mid-2013 and to finalize by December 2014. The new rule is expected to focus on stormwater discharges from newly developed and redeveloped, or post-construction, sites, such as subdivisions, roadways, industrial facilities, and commercial buildings or shopping centers.

BACKGROUND

Stormwater discharge systems are the pipes and sewer lines that carry rainwater or snow melt, but not domestic sanitary wastes, away from urban areas and commercial and industrial facilities. For many years the focus of the nation's water quality programs was on controlling pollutants associated with industrial *process* wastewaters and municipal sewage discharges. More diffuse and episodic discharges (such as rainfall runoff from farm lands and urban runoff) and discharges believed to be relatively uncontaminated received less attention from policymakers and regulators.

However, as the traditional sources of water pollution have become better controlled through laws and regulations, attention has increasingly focused on remaining problems that continue to prevent attainment of state and tribal water quality standards. Stormwater is one such source of pollution. For some time, it was generally believed that stormwater was largely clean, or uncontaminated.

However, studies have demonstrated that this type of discharge—from rainfall and snow melt—carries with it large amounts of organic and toxic pollutants that can harm water quality, including oil and grease, heavy metals, pesticides, soil, and sediment.

In urban areas, widespread residential and commercial development results in the removal of vegetation cover and building of impervious structures such as roads and parking lots. These activities may change natural drainage patterns in an area, causing higher runoff flows during wet weather events.

Urbanization and the human alteration of landscapes and land uses that is associated with it have resulted in the degradation of conditions in downstream waterbodies. States report that stormwater discharges, including urban runoff, industrial activity, construction, and mining, are a significant source of surface water quality problems today.

But control of stormwater discharges and other sources of wet weather pollution, including overflows from combined and separate sewer systems, is complicated because discharges generally are intermittent and are less amenable to "end of pipe" solutions than conventional industrial and municipal water pollution.

Recognition of the water quality problems of stormwater runoff led Congress in 1987, when it last comprehensively amended the Clean Water Act (CWA), to direct EPA to implement a specific permit program for stormwater discharges from industrial sources and municipalities (P.L. 100-4). Even before the 1987 amendments, the issue of how to regulate stormwater discharges had a lengthy history of regulatory proposals, delays, legal challenges, and court decisions. Still, EPA had been unable to devise a comprehensive and flexible administrative process for regulating stormwater discharges before requirements were legislated in 1987. In that legislation, Congress established a phased and tiered approach to permitting of stormwater discharges that fundamentally redesigned the CWA's approach to stormwater discharges.

Congress recognized that EPA's difficulties in addressing sources of stormwater stemmed in part from the large number of sources potentially subject to regulation, so the 1987 legislation adopted a procedure that would enable the major contributors of stormwater pollutants to be addressed first, and all remaining stormwater discharges in later phases.

EPA initially issued regulations to implement Congress's legislative mandate in 1990, utilizing a series of phased requirements.[1] Phase I applied to large dischargers: those associated with industrial activities, municipal separate storm sewer systems serving 100,000 people or more, and construction projects disturbing more than 5 acres. Smaller sources were slated for possible regulation under Phase II of the program (discussed below) and included cities and towns with separate storm sewer systems serving fewer than 100,000 people, commercial operations, and smaller construction projects. Stormwater requirements are one element of the comprehensive permit program, the National Pollutant Discharge Elimination System (NPDES), authorized in Section 402 of the act. Under the act, it is illegal to discharge pollutants from point sources (e.g., industrial plant pipes, sewage treatment plants, or storm sewers) into the nation's waters without an NPDES permit—permits are the fundamental compliance and enforcement mechanism of the law.

EPA manages the NPDES stormwater program in four states (Idaho, Massachusetts, New Hampshire, and New Mexico), plus the District of Columbia and most U.S. territories, and has delegated that authority to the remaining 46 states and the Virgin Islands.

An estimated 123,000 industrial facilities (twice the number of industrial sources subject to the base NPDES program) and 220 municipalities and counties were covered by the 1990 permit rules for Phase I of the program. The initial procedures and deadlines were complex and were made more confusing by subsequent deadline extensions.[2]

The 1987 CWA amendments directed delegated states (or EPA) to issue stormwater permits not later than four years after enactment of that legislation. This would have required permits to be issued by February 4, 1991, but this did not occur, in part because EPA's 1990 rule was issued 21 months after the statutory deadline. Regulated sources must comply with stormwater permits within three years of their issuance.

Permits require dischargers, at a minimum, to implement pollution prevention plans, although remediation or additional treatment of runoff may also be required. Permits issued to municipalities require cities to develop, implement, and enforce a stormwater management program that addresses key areas such as public education, eliminating illicit connections to storm sewers, good housekeeping of municipal operations, and control of erosion and sedimentation from construction sites.

Prior to implementation of the stormwater regulatory program, the universe of NPDES permittees nationwide was less than 70,000 industrial and municipal facilities.

The addition of stormwater permittees greatly expanded this regulatory program. EPA estimates that the total number of stormwater permittees at any one time exceeds half a million—thus, NPDES stormwater permittees outnumber wastewater permittees more than five-fold.

Industrial Facilities

Industries that manufacture, process, or store raw materials and which collect and convey stormwater associated with those activities were required to apply for an NPDES permit under the Phase I program. Several industries were specifically identified in EPA's 1990 regulation: mining operations; lumber and wood products; paper and allied products; printing, chemical products, paints, varnishes, and lacquers; stone, clay, glass, and concrete; metals; petroleum bulk terminals; hazardous waste treatment facilities; salvage operations; and powerplants.

Industrial facilities had several options to comply with these permit requirements. Chiefly, they could obtain either individual or group permits. Applications for *individual* facility permits were due to be submitted by October 1, 1992. For *group* permits (covering multiple facilities with similar stormwater discharges), a two-step process applied: submitting a list of facilities to be covered by September 30, 1991, and submitting more detailed information, such as sampling data on 10% of facilities in the group and a description of a stormwater management program, by October 1, 1992.[3]

EPA also provided a third option for industrial facilities, through a *general* permit procedure. A general permit is one that covers discharges from more than one facility, thus making the large number of stormwater permittees more manageable. Sources are only required to submit a Notice of Intent to be covered by a general permit, rather than the detailed application for an individual permit. EPA expected that general stormwater permits will make for a less costly and burdensome permitting process through less extensive testing and control requirements, as well as minimal monitoring and reporting. For most sources, general permits require preparation of a pollution prevention plan, and compliance with the plan six months later. EPA has issued general permits for stormwater discharges associated with industrial and construction activities that disturb 5 acres or more, which apply in the four states where EPA is the permitting authority for the stormwater program. Using the EPA general permit as a model, most other states that have been delegated permitting responsibility use similar general permits to reduce the administrative burden of the industrial stormwater permit program.

Since 1987, Congress has twice addressed the deadlines for stormwater permitting of industrial facilities. Congress first extended aspects of the deadlines for group applications by industrial facilities (P.L. 102-27, Dire Emergency Supplemental Appropriations Act of 1991), and in the 1991 Surface Transportation Act (P.L. 102-240), Congress clarified the deadlines applicable to industrial activities that are municipally owned or operated (such as airports or powerplants).

Municipalities

Phase I

Much of the controversy about stormwater requirements has focused on impacts on cities, not industrial sources. Municipalities with separate storm sewer systems (called MS4s) were subject to EPA's regulations under staggered deadlines based on the size of population served. In the 1990 Phase I regulations that apply to industrial activities, EPA also regulated discharges from medium-size and large cities (covering those with populations greater than 100,000 persons). The Phase 1 regulations are primarily application requirements that identify components that must be addressed in permit applications. The rules require large and medium MS4s to develop a stormwater management program, track and oversee industries facilities that are regulated under the stormwater program, conduct monitoring, and submit periodic reports. The regulations specified deadlines for these cities to provide regulators with information on legal authority over stormwater discharges and to provide detailed information on source identification and monitoring data. EPA identified 173 cities and parts of 47 urban counties as covered by Phase I.

Phase II

The 1987 CWA amendments exempted smaller cities (with populations of fewer than 100,000) from any stormwater permit requirements until October 1, 1992, and directed EPA to develop a suitable approach to address them under Phase II of the stormwater regulatory program. Because of problems in formulating a permitting strategy, EPA did not issue regulations by the 1992 deadline, nor did it meet the deadline in a one-year extension that Congress provided in P.L. 102- 580. In 1995, EPA convened an advisory committee of stakeholders to assist in developing rules by March 1, 1999, a deadline set in a judicial consent order in *Natural Resources Defense Council v. EPA* (Civ. No. 95-0634 PFL [DDC, Apr. 6, 1995]) that required EPA to clarify the scope of coverage and control mechanisms for the Phase II program. Based in part on extensive discussions with the stakeholder advisory committee and with another court-approved extension, EPA issued a final Phase II rule in 1999.[4] EPA estimated that the rule would make approximately 3,000 more river miles safe for boating annually and protect up to 500,000 people a year from illness due to swimming in contaminated waters.

The 1999 Phase II rule extended Phase I by requiring permits of two additional classes of dischargers on a nationwide basis: (1) operators of MS4s serving populations of less than 100,000 persons in urbanized areas as defined by the Bureau of the Census, and (2) operators of construction activities that disturb greater than 1 and less than 5 acres of land (larger construction sites are covered by the Phase I rules). Separate storm sewer systems such as those serving military bases, universities, large hospital or prison complexes, and highways are also included in the definition of small MS4. EPA estimated that 5,040 small cities are covered by Phase II, along with about 110,200 construction starts per year.[5]

Waivers from coverage are available both for small cities (those with fewer than 10,000 persons) and construction activities if the discharges are not causing water quality impairment. At the same time, additional small municipal systems and construction sites may be brought into the stormwater program on a case-by-case basis, if permitting authorities determine that they are significant contributors to water pollution. Under the 1999 rule, covered facilities were required to apply for NPDES permit coverage by March 2003 (most under a general rather than an individual permit) and implement stormwater management programs that include six minimum management controls that effectively reduce or prevent pollutant discharges into receiving waters, such as pollution prevention and eliminating illicit discharge connections for municipal operations. The rule also provided that municipally operated industrial activities not previously regulated were required to apply for permit coverage under the same schedule as other facilities covered by Phase II.

In the final Phase II rule, EPA attempted to balance statutory requirements for a nationally applicable program with sufficient administrative flexibility to focus on significant water quality impairments. For example, EPA encouraged permitting authorities to use general rather than individual permits for the majority of covered dischargers. The agency's decision to not include construction sites smaller than 1 acre was based on the belief that regulating the smallest of such sites would overwhelm the resources of permitting authorities and might not yield corresponding water quality benefits. Further, EPA modified the previous Phase I rule to exclude industrial facilities that have "no exposure" of their activities (such as raw materials) to stormwater, thus reducing coverage by an estimated 76,000 facilities that have no industrial stormwater discharges. These efforts to provide flexibility notwithstanding, many regulated entities continued to criticize the scope of the stormwater program, saying that EPA had greatly underestimated the cost of the Phase II rules (projected to be $297 million annually for small cities and $505 million annually for construction activities).

Cities of all sizes have complained about the costs and difficulties of complying with EPA's regulations, especially because there is no specific CWA grant or other type of assistance program to help pay for developing and implementing local stormwater programs. Many contend that cities already are burdened with numerous environmental compliance requirements and lack adequate resources to address stormwater controls in addition to drinking water, solid waste, wastewater treatment, and sludge disposal problems. Where cities need to construct or install technology to control stormwater discharges, the principal source of financial assistance is the CWA's state revolving fund (SRF) loan program that is administered by states. However, SRF assistance is not restricted to meeting just stormwater project needs, competition for available funds for all types of eligible projects is intense.[6]

Many municipal and industrial dischargers covered by the Phase I and Phase II programs have reached the end of their initial permit terms (NPDES permits are issued for five-year terms). For permit renewals, the agency is implementing a streamlined reapplication process that will not require the extensive information collection that characterized the first round of permitting.

Implementation of permits (i.e., translating permits into specific steps to manage stormwater runoff) is now the challenge for permitting authorities and permittees. According to a 2001 Government Accountability Office (GAO) report, local governments are primarily using best management practices (BMPs, sometimes called stormwater control measures, or SCMs) to prevent or slow stormwater from quickly reaching nearby waterbodies and degrading water quality, rather than requiring that stormwater be transported to treatment facilities.[7] BMPs include nonstructural measures to minimize contaminants getting into stormwater (e.g., street sweeping) and structural practices such as detention ponds to separate contaminants from stormwater. GAO criticized EPA for not establishing systematic efforts or measurable goals to evaluate the effectiveness of the program in reducing stormwater pollution or to determine its costs, which local governments have portrayed as high. In the 1999 rules, EPA set a goal of beginning to evaluate implementation of Phase II of the program in 2012.

In a 2007 report, GAO examined implementation of the stormwater regulatory program by municipalities.[8] GAO found that implementation of both Phases I and II has been slow: nearly 11% of communities were not permitted as of 2006; and even in communities with permits, delays occurred due to litigation or other disputes. Thus, GAO reported that because many communities are still in the early stages of implementation, it is too early to determine the overall program burden. While EPA's regulations provide flexibility, which could limit program burden, increased burden could result if communities are required by states or EPA to expand stormwater management activities or meet more stringent specific permit conditions in the future. GAO found that EPA is not collecting complete and consistent cost and other data, which hampers assessment of program burden.

CONGRESSIONAL INTEREST

Prior to issuance of the final Phase II rule in 1999, Congress included language in EPA's FY2000 appropriation bill (P.L. 106-74) directing the agency not to issue the final rule before submitting a detailed impact analysis to Congress. To meet a court-ordered deadline for the regulation, EPA released the report concurrently with the Phase II rule.[9] In the 106th Congress, legislation was introduced to exempt construction sites of less than 5 acres and certain above-ground drainage ditches from stormwater permitting requirements (S. 2139/H.R. 3625). At an October 1999 Senate hearing, EPA witnesses opposed the bill, saying that above-ground drainage ditches and small construction sites are significant sources of water pollution and thus should be subject to stormwater management requirements. No further action occurred. In response to concerns about program impacts and costs, the 107th Congress enacted legislation allowing states to use Section 319 grant funds, which are used for projects to manage nonpoint sources of water pollution, for projects or activities related to developing and implementing a Phase II stormwater program (§301 of P.L. 107-303). This authority only applied to Section 319 funds in FY2003. Legislation to extend this authority beyond FY2003 was introduced in the 108th Congress, but was not enacted (S. 1716/H.R. 3528).

Oil and Gas Facilities

As the March 2003 Phase II deadline approached (affecting small municipalities and construction sites), EPA proposed a two-year extension of the rule for small oil and gas exploration and production facility construction sites to allow the agency to assess the rule's economic impact on that industry. EPA had initially assumed that most oil and gas facilities would be smaller than one acre in size and thus excluded from Phase II rules, but newer data indicated that up to 30,000 new sites per year would be of sizes subject to the rule. In March 2005 EPA extended the exemption until June 2006 for further study and said it would issue a specific rule for small oil and gas construction sites by that date. The postponement did not affect other industries, construction sites, or small cities covered by the 1999 rule. Under the 1987 amendments to the CWA, the *operations* of facilities involved in oil and gas exploration and production generally were exempted from compliance with stormwater runoff regulations (so long as the runoff is uncontaminated by pollutants), but the *construction* of associated facilities was not.

Omnibus energy legislation enacted in the 109th Congress (P.L. 109-58, the Energy Policy Act of 2005) included a provision addressing this issue. Section 323 amends the CWA to specifically include construction activities at *all* oil and gas development and production sites, regardless of size (including sites larger than 5 acres, previously covered by Phase I), in the law's general statutory exemption for oil and gas facilities from stormwater rules. Its intention was to exempt from the CWA all uncontaminated stormwater discharges that occur while setting up drilling operations.

Oil and gas officials, who supported the provision, said that the existing EPA stormwater rules create time-consuming permitting requirements, even though the short construction period for drilling sites carries little potential for stormwater runoff pollution. Opponents argued that the provision did not belong in the omnibus energy legislation and that there is no evidence that construction at oil and gas sites causes less pollution than other construction activities, which are regulated under EPA's stormwater program.

EPA promulgated a rule to implement Section 323 in 2006.[10] The rule was criticized by some interest groups and Members of Congress who argued that EPA had exceeded its authority by broadly defining the scope of contamination that is exempted by the rule beyond the statutory language to also include stormwater discharges contaminated solely with sediment. In May 2008, a federal court held that the rule is arbitrary and capricious, and it vacated the rule.[11] EPA petitioned the court to rehear the case, but the request was denied—thus, the exemption is no longer in effect. EPA intends to issue a revised rule that would remove the 2006 rule from the Code of Federal Regulations consistent with the court vacatur and codify the statutory exemption in P.L. 109-58, but the agency has not announced a specific schedule for doing so. Legislation to repeal Section 323 was introduced in the 109th Congress (H.R. 4541), but no further action occurred.

In the 111th Congress, legislation to repeal the exemption that was enacted in P.L. 109-158 passed the House on July 30, 2010 (the provision was Section 728 of H.R. 3534, the Consolidated Land, Energy, and Aquatic Resources Act), but there was no further action on this bill.

Stormwater Management at Federal Facilities

Congress often looks to federal agencies to lead or test new policy approaches, a fact reflected in legislation passed in 2007. Section 438 of P.L. 110-140, the Energy Independence and Security Act (EISA), requires federal agencies to implement strict stormwater runoff requirements for development or redevelopment projects involving a federal facility in order to reduce stormwater runoff and associated pollutant loadings to water resources. The legislation requires agencies to use site planning, construction, and other strategies to maintain or restore, to the maximum extent technically feasible, the predevelopment hydrology of the property.

To assist agencies in meeting these requirements, EPA issued technical guidance.[12] The guidance provides two options for meeting the performance objective of preserving or restoring the hydrology of a site: retaining the 95th percentile rainfall event (i.e., managing rainfall on-site for storm events whose precipitation total is less than or equal to 95% of all storm events over a given period of record), or site-specific hydrologic analysis (i.e., using site-specific analysis to determine predevelopment runoff conditions). According to the guidance, using a performance-based approach rather than prescriptive requirements is intended to give site designers maximum flexibility in selecting appropriate control practices. Issuance of the guidance also fulfilled an element of an October 2009 executive order that formally assigned to EPA the responsibility to issue the Section 438 guidance, in coordination with other agencies, and to do so by December 5, 2009.[13]

In December 2010 Congress passed legislation requiring federal agencies to pay local fees for treating and managing stormwater runoff. The legislation amends CWA Section 313, which requires federal agencies to comply with all federal, state, and local water pollution control requirements as nongovernmental entities, including the payment of reasonable service charges. The issue emerged earlier in 2010 when several federal agencies announced that they would not pay stormwater fees assessed by the District of

Columbia, claiming that the fees amounted to a tax that the agencies were not required to pay, because the waiver of sovereign immunity in Section 313 applies to fees and charges, but not a tax. The legislation was intended to clarify uncertainty over whether federal agencies must pay local stormwater fees. President Obama signed the legislation in January 2011 (P.L. 111-378).

A continuing aspect of the issue of interest is the scope of P.L. 111-378 and whether it requires the government to pay local stormwater fees retroactively. After the Bonneville Power Administration had objected to paying retroactive stormwater fees imposed by two Washington localities following passage of the federal facility amendment, the matter ended up in federal court. The government took the position in the litigation, as well as in similar cases in Georgia and Wisconsin, that the legislative change amounted to a redefinition of "service charges," instead of a clarification of Congress's original intent, and would only apply prospectively. In May 2012, a federal district court rejected the government's position and held that the CWA amendment was merely a clarification of the statute and thus is entitled to retroactive effect.[14]

Recent Developments: NRC Report and Potential Regulatory Changes

In 2006 EPA requested the National Research Council of the National Academy of Sciences (NRC) to conduct a review of the existing stormwater regulatory program. The resulting report, issued in 2009, called for major changes to EPA's stormwater control program that would focus on the flow volume of stormwater runoff instead of just its pollutant load. The committee observed that—

> stormwater discharges would ideally be regulated through direct controls on land use, strict limits on both the quantity and quality of stormwater runoff into surface waters, and rigorous monitoring of adjacent waterbodies to ensure that they are not degraded by stormwater discharge.... Presently, however, the regulation of stormwater is hampered by its association with a statute that focuses primarily on specific pollutants and ignores the volume of discharges.[15]

The NRC report recommended that EPA adopt a watershed-based permitting system encompassing all discharges—stormwater and wastewater—that could affect waterways in a particular drainage basin, rather than individual permits that do not account for cumulative conditions from multiple sources in the same watershed. Under the proposed watershed permitting strategy, responsibility to implement watershed-based permits and control all types of municipal, industrial, and construction stormwater discharges would reside with MS4 permittees. The report criticized EPA's current approach, which leaves much discretion to regulated entities to set their own standards through stormwater management plans and to self-monitor. As a result, enforcement is difficult and variable, and information to assess the water quality benefits of the regulatory program is limited. The report also noted that adequate resources, including new levels of public funds, will likely be required to operate a more comprehensive and effective stormwater permitting program.

In response, EPA initiated information-gathering and public dialogue activities as a prelude to possible regulatory changes that would respond to the NRC's criticism of inconsistency in stormwater requirements nationally and embrace the report's recommendation to adequately control all sources of stormwater discharge that contribute to waterbody impairment. EPA proposed to collect data from MS4s, states, and industry entities involved in developing or redeveloping sites on the scope of the current regulatory program and management practices, as well as information on control, pollution prevention technologies, and BMPs applied to stormwater discharges from newly developed and redeveloped sites.[16]

EPA is exploring regulatory options that would strengthen the regulatory program, including establishing specific post-construction requirements for stormwater discharges from new development and redevelopment, which currently are not regulated. While MS4s are required to address stormwater discharges from new development and redevelopment in their management plans, EPA rules do not include specific management practices or standards to be implemented.

Other options that EPA is considering include expanding the area defined as MS4s to include rapidly developing areas, devising a single set of consistent regulations for all MS4s, and requiring MS4s to address stormwater discharges in areas of existing development through retrofit practices. The rule is expected to focus on stormwater discharges from developed, or post-construction, sites, such as subdivisions, roadways, industrial facilities, and commercial buildings or shopping centers, and to seek to ensure that even after development projects are completed, runoff levels from sites are equivalent to pre-construction hydrology. As part of the national post-construction stormwater rule, EPA has committed to consider supplemental provisions that would apply only to the Chesapeake Bay watershed, a region where municipal stormwater discharges are a significant cause of water quality impairment and are one of the only sources of pollutants with increasing loads to the Bay and its tributaries. EPA officials note that a number of states have been developing their own stormwater management programs, particularly in the Northeast, where lawsuits have pushed regulators, and in some high-precipitation Northwest states.

In early 2010 EPA held a series of listening sessions across the country as part of a process seeking public comments on potential considerations for regulatory changes.[17] The agency also sent survey questionnaires to property owners and developers, municipal sewer system authorities, state regulators, and EPA regional offices to obtain their input. Some industry groups reportedly criticized possible expansion of the current program, saying that EPA's authority to regulate stormwater does not extend to regulation of post-construction discharges. Some states also said that EPA lacks the technical knowledge to regulate stormwater across the nation, while states with comprehensive regulations, such as Florida and Maryland, demonstrate that regulation is best done at the state and local level, because of locational differences in stormwater discharges. A number of commenters urged EPA to ensure that performance standards designed to reduce storm runoff be flexible so that communities can create requirements appropriate to their stormwater needs. Cost is a key issue raised by states and municipalities that are concerned about the possibility of a mandatory retrofit requirements that would impose a significant economic burden on cities.[18]

EPA had been considering possible changes to the 1990 and 1999 stormwater rules since 1999. But the effort progressed slowly until a deadline was set in a 2010 settlement with the Chesapeake Bay Foundation.[19] Under that settlement, EPA was expected to issue a proposed rule in September 2011 and to promulgate a final rule by November 2012. The parties later agreed to extend the deadline for a proposed rule to April 27, 2012. However, in July, EPA announced on its website that it will propose its post-construction rule in June 2013, almost two years beyond the original deadline mandated under the settlement agreement, and issue a final rule by December 2014. The rule has reportedly been delayed while the agency considers costs and benefits of its proposal and ways to incorporate flexibility, such as possibly including lengthy implementation plans for retrofit projects and allowing states with equivalent stormwater programs to regulate in lieu of EPA.

End Notes

[1] U.S. Environmental Protection Agency, "National Pollutant Discharge Elimination System Application Regulations for Storm Water Discharges," 55 Federal Register 47990-48091, November 16, 1990.

[2] Moreover, the 1990 regulations themselves were challenged by an environmental group, the Natural Resources Defense Council. In 1992, a federal appeals court ruled that EPA had failed to meet certain deadlines specified in the 1987 legislation and had been improper in exempting from regulation light industry and construction sites that affect less than 5 acres of land. Natural Resources Defense Council v. EPA, 966 F.2d 1292 (9th Cir. 1992).

[3] The same deadlines also applied to industrial activities owned or operated by municipalities with a population of 250,000 or more. For industrial activities owned or operated by municipalities with populations of more than 100,000 but less than 250,000, the respective deadlines were May 18, 1992, and May 17, 1993. Certain government-owned or - operated activities (airports, powerplants, and uncontrolled sanitary landfills) also were subject to the May 1992 and May 1993 deadlines, even if the unit of government has a population of less than 100,000.

[4] U.S. Environmental Protection Agency, "National Pollutant Discharge Elimination System—Regulations for Revision of the Water Pollution Control Program for Storm Water Discharges," 64 Federal Register 68721-68851, December 8, 1999.

[5] Although precise numbers are not available, EPA now estimates that the number of regulated MS4s is about 7,000 (including 1,000 under Phase I and 6,000 under Phase II). The number of industrial permittees is estimated to be around 100,000. The construction site portion of the program each year covers about 200,000 permittees under Phase I (5 acres or greater) and another 200,000 under Phase II (1 to 5 acres). See National Research Council of the National Academy of Sciences, Water Science and Technology Board, Urban Stormwater Management in the United States, The National Academies Press, Washington, DC, 2009, p. 36.

[6] For additional information, see CRS Report 98-323, Wastewater Treatment: Overview and Background, by Claudia Copeland.

[7] U.S Government Accountability Office, Water Quality, Better Data and Evaluation of Urban Runoff Programs Needed to Assess Effectiveness, GAO-01-679, June 2001.

[8] U.S. Government Accountability Office, Clean Water, Further Implementation and Better Cost Data Needed to Determine Impact of EPA's Storm Water Program on Communities, GAO-07-479, May 2007.

[9] U.S. Environmental Protection Agency, Office of Water, Report to Congress on the Phase II Storm Water Regulations, EPA 833-R-99-001, September 1999, http://www.epa.gov/npdes/pubs/ReptoCong_PhII_SWR.pdf.

[10] U.S. Environmental Protection Agency, "Amendments to the National Pollutant Discharges Elimination System (NPDES) Regulations for Storm Water Discharges Associated with Oil and Gas Exploration, Production, Processing, or Treatment Operations, or Transmission Facilities," 71 Federal Register 33628-33640, June 12, 2006.

[11] Natural Resources Defense Council v. U.S. Environmental Protection Agency, 526 F.3d 591 (9th Cir. 2008).

[12] U.S. Environmental Protection Agency, Office of Water, Technical Guidance on Implemementing the Stormwater Runoff Requirements for Federal Projects Under Section 438 of the Energy Independence and Security Act, EPA 841- B-09-001, December 2009, http://www.epa.gov/owow/NPS/lid/section438/pdf/final_sec438_eisa.pdf.

[13] Executive Order 13514, "Federal Leadership in Environmental Energy and Economic Performance," 71 Federal Register 52117-52127, October 8, 2009.

[14] United States v. Renton, W.D. Wash., No. C11-1156JLR, May 25, 2012.

[15] Supra note 5, p. 3.

[16] U.S. Environmental Protection Agency, "Agency Information Collection Activities; Proposed Collection; Comment Request; Stormwater Management Including Discharges from Newly Developed and Redeveloped Sites; EPA ICR No. 2366.01, OMB Control No. 2040-NEW," 74 Federal Register 56191-56193, October 30, 2009.

[17] U.S. Environmental Protection Agency, "Stakeholder Input; Stormwater Management Including Discharges from New Development and Redevelopment," 74 Federal Register 68617-68622, December 28, 2009.

[18] Examples of urban retrofits include breaking up concrete and installing stones, thereby replacing an impermeable surface with a permeable one that will absorb runoff; and adding vegetation and trees to parking lots.

[19] In addition to responding to the 2009 NRC report, EPA's development of the stormwater rule is partly driven by the terms of a 2010 settlement agreement with environmental advocacy groups, which forced the agency to issue a Total Maximum Daily Load (TMDL) pollution control plan for the Chesapeake Bay. Fowler v. EPA, D.D.C. No. 1:09-cv-5, May 11, 2010.

INDEX

D

E

F

G

H

I

J

L

M

Q

R

S

T

U

V

W

Y